ASTRO NAVIGATION

by Pocket Computer

ASTRO NAVIGATION

by *Pocket Computer*

M. J. Harris

ADLARD COLES
8 Grafton Street, London W1

Loading programs from a cassette tape or floppy disk is many times quicker and more certain than loading them by typing in the listings. For information on the availability of programs in this form please send a stamped and self-addressed envelope to: Adlard Coles, 8 Grafton Street, London W1X 3LA.

Adlard Coles
William Collins Sons & Co. Ltd
8 Grafton Street, London W1X 3LA

First published in Great Britain by
Adlard Coles 1989

British Library Cataloguing in Publication Data

Harris, M. J.
Astro navigation by pocket computer
1. Nautical astronomy. Data processing
I. Title
623.89

ISBN 0-229-11846-1

Printed and bound in Great Britain by
Hartnolls Ltd, Bodmin

Contents

Acknowledgements   ix
Preface   xi
Using this Book   xii

1 The Background to Small Boat Astro Navigation
Geographical Positions   2
Measuring Angles and Finding Latitudes   3
The Longitude Prize   6
19th-Century Software Innovation   7
Mathematical Difficulties   8
The Last 40 Years   9
For the Future   10

2 Tools and Skills
The Watch   11
The Sextant   13
Mechanical Sextant Adjustments   14
   Perpendicularity error   14
   Side error   14
   Index error   15
Mathematical Adjustments   16
Difficulties Caused by the Weather   16
The Computer   18
Pocket Computers for Use at Sea   18
Typing in Program Instructions   20
Star Finders   21

3 Terms and Symbols Explained

Latitude and Longitude Coordinates (LAT and LNG)   22

The Nautical Mile   22

Declination (DEC)   25

Hour Angles   26

   Sidereal Hour Angle (SHA)   27

   Greenwich Hour Angle (GHA)   28

   Local Hour Angle (LHA)   28

Greenwich Mean Time   28

   Greenwich Date   29

Using the Computer   29

   Variables   31

   Line Numbers   32

   Working with Angles   32

Instructions in BASIC   33

Memory Space   36

4 Taking Sextant Angles

Vertical Sextant Angles Used in Coastal Navigation   37

Sextant Angles in Astro Navigation   37

The Menu   40

Computing with Sextant Angles   41

Index Error   42

Corrections for Dip and Refraction   42

The Planets   44

The Sun and Moon   45

   Parallax Corrections for the Moon   46

Overall Accuracy   48

BASIC Listings   48

   (1) The Menu   48

   (2) Sextant Subroutine   48

   (3) Conversion of Degrees, Minutes and Seconds
       to Decimal Form   49

5 Sight Reduction

Predicting Positions of Objects in the Sky 50
The Connection between Predicted and Observed
  Altitudes 52
Programming for Sight Reduction 52
BASIC Listings 57
  (4) Control Program 57
  (5) Loading LAT and LNG 57
  (6) DEC and GHA Loading 58
  (7) Removing Excess Multiples of 360 Degrees 58
  (8) Sight Reduction 58

6 Adding Some Ephemeris

Handling Time 59
Control Program Adjustments 61
Star Ephemeris 62
Sun Ephemeris 62
Moon Ephemeris 63
Ephemeris from *Compact Data for Navigation and
  Astronomy* 63
Using the Coefficients 66
BASIC Listings 67
  Control Program Changes 67
  (9) Greenwich Date and Time 67
  (10) GHA Aries 68
  (11) Star DEC and SHA Loading 68
  (12) Sun DEC and GHA 69
  (13) Moon DEC and GHA 69
  (14a) Loading Routine for Compact Data Coefficients 71
  (14b) Compact Data Method for Planet DEC and GHA 71

7 Latitude and Longitude Fixes

Transferred Positions 73
LAT/LNG Fixes by Computer 75

Living and Working with Errors   75
Plotting versus Chartwork   78
Fixes from more than 2 Position Lines   78
BASIC Listings   79
   (15) Control Program Additions and Extension   79
   (16) LAT/LNG Fix   79

8 Emergency Astro Navigation
Loss of the Compass   82
Loss of the Computer   83
  Sight Planning   84
  High Altitude Sights   84
  Meridian Altitudes   84
Loss of the Sextant   87
  Horizon Sights   88
Loss of the Watch   88
Loss of Everything   89

Appendices

1 Commands used in other forms of BASIC
DEG and DMS   90
DEG and RAD   90
DEF and FN   91

2 Annual Updates   91

3 Allocation of Variables   92

4 Subroutine Mapping   94

5 Program Testing – Getting the Bugs Out   94

Conclusion   97

Bibliography   98

Acknowledgements

Writing this book would not have been possible without much help and encouragement from many people. I am especially grateful to Di, my wife, for her help with the illustrations and continued help with the text, and to Lucy for testing both the programs and computers.

I am also sincerely thankful for all the help and advice that the following people have given when it was most needed: Jimmy and Ann Griffin, Jim and Dorothy Scott, Roger Stoyl, Barbara and John Tappin, Gerts and Vivien Webster, Dr B. D. Yallop. Also to Aquaman (UK) Ltd, who supplied plastic bags, eminently suitable for protecting pocket computers at sea.

Preface

Small boat sailors have always prized astro navigation as a self-reliant system for fixing positions in mid-ocean. As beacons it uses the sun, moon, stars and planets, and unlike the modern all-electronic 'black box' systems is independent of the vagaries of international politics or the success of the space shuttle.

Getting the most out of the system requires finding accurate solutions to some mathematical relationships, and at sea this would be especially difficult were there not some method for simplifying the working. Traditional methods have heavily depended upon tables, but in recent years pocket calculators have been shown to play an important part in making the workings both faster and more convenient.

Now the trend continues as computers have become available with increasing reliability and ever decreasing size and power requirements. As prices continue to tumble and capabilities increase, pocket computers can provide small boat skippers with the means of fixing positions from the objects in the sky, as quickly as by using radio direction finders and almost as quickly as hand bearing fixes from visual objects.

Working out your own astro navigation programs from scratch can be an absorbing occupation but one that will consume a considerable amount of your time. The alternatives are to buy a ready-programmed dedicated navigational computer or to use a ready-published program, but here there are still difficulties. Ready-programmed computers are always less easily available than the general purpose computers that can be obtained at a fraction of the price from discount stores anywhere in the world. Furthermore, their programs are kept secret, which does little to increase the user's understanding of astro navigation, and there is often little scope for using the computer for other applications. Nor is buying published programs without its difficulties; if you don't have the particular IBL Spectrobeeb, XF 720 or Lemon for which it was written, you will probably have to rewrite it for your own computer.

This book contains a set of astro navigational programs that are an attempt to overcome these problems. They are written in BASIC, a language widely taught in schools and acceptable to most personal and pocket computers (although different machines use it in slightly different forms). However, the programs listed here tackle these difficulties by using only the simplest of BASIC instructions – those that are common to just about all versions. Originally, they were written for the Casio FX-720P, FX-730P, Sharp PC-1247 and PC-1403. There are also many other machines on which the programs will run with few difficulties but it is still possible that particular machines will have slightly differing requirements. In these cases the notes in Chapter 3 give the intentions behind the program statements used here, and these can be compared with similar statements appearing in the computer manual to show what changes are needed.

USING THIS BOOK
After the introductory chapters, Chapter 4 develops a program for correcting sextant angles. Particularly if you are a beginner to astro navigation or computing, it is a good idea to start by loading and testing this one. At the end of each of Chapters 4 to 6 you are left with a useful piece of program that can stand alone or be further developed in the following chapter.

It is almost certain that your particular computer will have far greater capabilities than those required by the programs listed here. This is because the programs are intended simply to provide a versatile framework for a working system. So, if your computer can print lower-case letters, has more useful mathematical functions, more memory, a larger display screen or some graphics capability, the possibilities for improving the clarity and convenience of the programs are endless. The programs offered are bare bones for you to tinker with and change as you see fit.

Finally, the last chapter includes a brief introduction to emergency astro navigation and some procedures you might consider in the event of equipment failure. Again, it is intended as a starting point from which to begin building your repertoire of possible techniques, rather than a definitive list of all possibilities.

Through this book I hope to give you an extra interest not just in pocket computers, which are simply a tool for the job, but in perfecting a familiarity with the sky as an aid to both routine and emergency navigation. It is an interest you can practise not only at sea but also at home or indeed at any time that you can see the sky.

1. The background to small boat astro navigation

As land animals we are born with an ability to form mental pictures of the surroundings in which we live. We have no difficulty in constructing vastly complex and very personal maps of where we live, go to work, visit our friends or go shopping. In a small boat in mid-ocean these skills go by the board if we can no longer recognize the signposts. It is not that they are no longer there but, unlike pigeons, dolphins, eels or salmon, we need to learn how to interpret what we see.

Finding our way about the earth from the 'map' of stars in the sky would be so much easier if only the earth would stop rotating. Then we would be able to name any position on the earth from the name of the star immediately above it. Going from one place to the next would simply mean heading for the star or position in the sky that we knew to be above it. Distances and directions on the ground would be in proportion to those in the sky.

The idea of stopping the earth perhaps sounds ridiculous; it would certainly cause a few problems. As most people are aware its rotation causes the stars to move from east to west across the sky each night. Nonetheless there is at least one star, the pole star, which is fixed above, and named after, the earth's north pole. In the southern hemisphere, too, although there is no similar star, the constellation of the Southern Cross does point to a corresponding position in the sky. Furthermore, even with a rotating earth, stars that appear overhead in one latitude will never appear overhead in any other latitude. So, if you can see Mintaka, a belt star in the constellation of Orion, overhead, you must be on the equator. If it is Eltanin, you are on the same latitude as London, and if it's Arcturus, your latitude is that of Hawaii and Cuba.

GEOGRAPHICAL POSITIONS
This idea of there being a one-to-one relationship between positions in the sky and those on the earth is important in astro navigation. Earthly positions of celestial objects are sometimes referred to as their *geographical positions*, and apart from the polar positions they are constantly changing. One of the first recorded uses of this idea was made by Eratosthenes in the 3rd century BC. Many around him still thought the earth to be flat, but he was a great believer in a round earth, and made

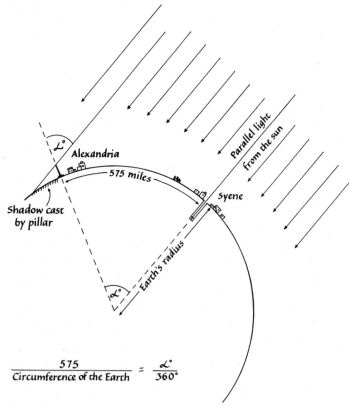

$$\frac{575}{\text{Circumference of the Earth}} = \frac{\alpha°}{360°}$$

Fig. 1.1 Eratosthenes' measurement of the circumference of the Earth

an attempt to measure its circumference. His equipment was little more than a good deal of local knowledge and observation. He knew that in Syene, a city on the River Nile (the present-day Aswan) there was a very unusual well. This well was special because for a brief moment on one particular day of the year the sun would be overhead and shining directly down into it. This moment was at noon on midsummer's day. At the same time in Alexandria, some 575 miles to the north, the sun was not vertically overhead. He was able to calculate the angle of the sun here by measuring the length along the ground of a shadow made by a tower of known height. This provided him with enough information to apply some geometry and estimate the size of the earth (Fig. 1.1). Presumably he estimated the distance between the two cities from the time it took people or camels to travel between them, but his estimate of the earth's circumference was amazingly accurate – just 15 per cent greater than present-day estimates.

MEASURING ANGLES AND FINDING LATITUDES

The sun and the pole star were probably the first objects to be routinely used by astro navigators. In the 9th century, northern sailors are known to have made use of the pole star, though at that time it was 7 degrees away from the true polar position. Nonetheless, it was still a useful direction indicator. Also important, especially in later centuries, was that its height above the horizon could be used to give a direct indication of latitude. Were you to stand at the north pole, you would find the pole star to be at your *zenith*, which is the point in the sky directly overhead. The angle between your zenith and the horizon is 90 degrees, so this corresponds with your earthly latitude, which is also 90 degrees north. As you travel further south the pole star falls lower in the sky until, when you pass the equator, it drops below the horizon. Though in the southern hemisphere there is no such conveniently situated pole star, the relative positions of nearby constellations can still be used to indicate where such a star would be if it existed. As you go further south this point in the sky rises and can similarly be used to find your latitude.

Measurement of latitude formed an important part of early navigation in many cultures. One of the simplest instruments was that reputed to have been used by the Polynesians sailing north between Tahiti and Hawaii.

Hawaii is a mere speck amidst a vast ocean and to miss it would have been disastrous. Beyond, there is 1800 miles of ocean to cross before hitting Alaska or the Aleutian Islands. The instrument of navigation consisted of a hollowed *cala-bash,* or gourd. The top was cut off and four holes were pierced at equal intervals around the circumference, each the same distance below the rim. Prevailing winds in Tahiti meant that vessels always left on a starboard tack and steered north-east. Some time after crossing the equator the pole star would become visible, and the time had come to use the calabash. It was filled with water just up to the level of the four holes and the navigator would view the star through one of the holes. If he could see it level with the opposite rim the vessel had reached the latitude of Hawaii and it was time to sail west. Each following night the latitude would be checked in this way until Hawaii appeared.

Unfortunately, not all experts have agreed that the particular calabash from which these ideas were drawn was in fact a nautical intrument. Others have suggested that it was a container for the chief's plumes and feathers during the voyage. Whatever it was used for, the navigational principles remain sound. The water provided an artificial horizon plane, set at the level of the holes. The angle between the surface of the water, and a line passing through one hole and touching the opposite rim, corresponded to the latitude of Hawaii. If this device was ever used for navigation then it would certainly have needed much practice to use it effectively and without getting an eyeful of water.

Even under ideal conditions it is hard to imagine how some early instruments for measuring angles could have given results accurate to within much less than 1 degree. If seas were bad then the results would have been very much worse and, since 1 degree of error represents 60 miles, it is easy to see the need for some improvement to the equipment – not only for fixing positions at sea more accurately, but also to produce better charts. There was little point in knowing your exact latitude and longitude if you did not know the position of the reef you were trying to avoid. Stemming from the need to produce better charts, most significant developments up to the mid-18th century were attempts at producing improved methods for measuring angles.

The *cross staff* was an early example and consisted of a horizontal wooden batten with a scale, which was a sliding fit in a shorter vertical cross-member. This device was often used for taking sun sights. In use the navigator held his eye at the end of the long batten and sighted the horizon. The cross batten was then slid until its tip just coincided with the sun. The altitude could then be read off from the scale.

Later came the *backstaff*, so named because the navigator stood with his back to the sun and observed the shadow made upon the scale. The working accuracies of these instruments were variable, but usually not much better than one-sixth of a degree and certainly a lot worse if conditions were bad. Efforts to improve upon this resulted in all manner of different staffs and quadrants being developed. The *quadrant* had an engraved degree scale and the navigator sighted along one edge. Some used the horizon as reference whilst others used a plumb line. A variety of designs fell into and out of favour, but the development of the *mirror sextant* had to wait for the beginning of the 18th century. This forerunner of today's sextants was developed by at least two people working independently, and became widely adopted to the exclusion of all else.

Before this time, no one sailed to a foreign port by the most direct route even if the prevailing winds were favourable to it. Instead it was safer to sail north or south until the latitude of the port was reached, and then to sail east or west on a constant latitude in order to reach it, rather in the way it was thought that the Polynesians used to find Hawaii. Even when latitude was quite uncertain other devices were used. For example, the traditional advice given to masters heading for the West Indies was: 'South until the butter melts, then west.'

The most popular method for finding a vessel's latitude was by measuring the angular height of the sun above the horizon at local noon. This was when it was the highest in the sky, and the navigator needed to know the latitude of its geographical position on that particular day. The handy device that was used here was the seaman's astrolabe, a multi-purpose ready reckoning instrument used over many centuries. It consisted of a suspended disc engraved with scales and a pointer. Just a few lines of arithmetic were all that were needed to fix a vessel's latitude (see Chapter 8). This procedure was common as a traditional daily part of a ship's routine, but for trading vessels to compete in a commercial world what was needed was a measurement of longitude.

THE LONGITUDE PRIZE

By observing the sun throughout the day navigators were able to estimate the time of local noon, no matter where they were in the world. Because the earth rotates once in 24 hours, the time of noon is shifted as one goes east or west. To find their longitude, what was needed was a measure of the time difference between their local noon and that of, say, their home port. This difference would have been proportional to the distance east or west that had been sailed.

Land-based clocks were able to maintain reasonable precision, but at sea things were different and the motion of the ship would make them speed up or slow down. These problems led to several countries making extensive efforts to find solutions, and in 1713 the British government set up the 'Longitude Board' and offered a prize of £20,000.

Lighthouses that flashed time signals onto the clouds and observations of eclipses of the moons of Jupiter were among the proposals that received serious attention but there were two other more practical contenders. One involved using the moon, and the other the production of a chronometer able to maintain its accuracy over long periods at sea.

The moon moves very rapidly across the sky and its movements are mathematically predictable. It had been suggested that its movement against the background of slower-moving stars could form a basis for measuring time. The theory was good, but there were two main problems. Firstly, for the method to be useful, it was necessary to measure the angle between the moon and a star to within a minute, and with the equipment of the day this was impractical. Secondly, the finer details of the moon's path are complex and the necessary observational and mathematical work had still to be completed. These problems had been recognized for some time and had led Charles II to set up the Royal Observatory at Greenwich, the site later to become the reference from which longitude was measured.

Eventually the major difficulties of measuring time by the lunar distance method were overcome, and amongst other famous navigators, Captain Cook was one who used it extensively. However, the calculations involved were extremely lengthy and needed meticulous attention to detail. This is one reason why the method has fallen into disuse. No doubt Captain Cook would have been pleased to have a pocket computer aboard.

The production of a chronometer suitable for determining longitude was the other approach. To win the longitude prize the chronometer needed to maintain its accuracy to within three seconds after a period of six weeks at sea. In 1728 John Harrison began work on the project. Six years later he produced a pendulum clock which was tested aboard a naval vessel and found to meet the specification. He was not given the prize but went on to produce other models including his most famous number 4, which had a balance wheel and was spring-driven. Eventually, after a long battle with the board and the intervention of the king, he was given the prize at the age of eighty-three.

19TH-CENTURY SOFTWARE INNOVATION

The 'hardware', or the tools used by astro navigators today, have changed little from those in use in the 1800s. Techniques for taking sights are still the same but the methods of working out the results are radically different and were introduced in the last century.

An American captain, Thomas Sumner, came up with the idea that a single sight could be used to determine a position line, along which the vessel was expected to lie. Like so many apparently innovative ideas, this one was not new. It was just that no one had thought about things in quite the same way before. Look back to the diagram showing how Eratosthenes measured the size of the earth (Fig. 1.1). The sun sight he took at Alexandria gave the same angle that he would have measured had he been at any point along a circle with a radius of 575 miles and centred upon the well in Syene. This was his position circle but his sun sight did not say where he was upon it.

Forty years later a French admiral, Marcq Saint-Hilaire, con-
tributed a general system for calculating position lines. It
formed the foundation of most methods in use today, including those used later in Chapter 5. It became possible to work
sights of the sun, moon, planets or stars, regardless of whether
they were to the north or south of the observer. Plotting the
results from a single sight upon a chart would produce a position line, whilst a pair of sights with intersecting lines fixed a
vessel's position. Chapter 7 gives a simple method for computing the fix without plotting.

MATHEMATICAL DIFFICULTIES

Now astro navigation was a position-finding system that could
fix the position of a vessel to within a couple of miles, or better
if conditions were good. It could be used at any time when the
horizon was visible together with the sun, the moon, or any of
the more conspicuous stars or planets. But there were still
difficulties, for what had been gained in versatility had been
lost in simplicity.

To use the system, navigators had to be equipped with astronomical data giving the predicted positions of all objects they
expected to take sights of. When they had taken a sight the
working that followed involved combining this positional data
with their approximate latitude and longitude and the sextant
reading. The result was a position line, but working it out
involved finding the solution to some problems in spherical
trigonometry and a considerable amount of time-consuming
attention to details. This is not the kind of thing that comes
easily when working under pressure in bad weather – yet
under these conditions it could be especially important to
know your position with absolute certainty.

Making these techniques workable at sea has been a preoc-
cupation for generations of astronomers and mathematicians.
Not mathematicians but seamen had to be able to work large
figures through some complex mathematics and be certain of
the answers. There was also the difficulty of finding the predicted positions of sun, moon, stars or planets at the time of the
sights taken.

Diagrams, and more particularly tables, had been successfully used to provide ready solutions to many mathematical

problems in the past. Tables were already used for tidal prediction, calculating interest rates, and many other mathematical jobs, so they provided an obvious way of transforming the apparently intractable mathematics of astro navigation into a more practical technique. Anyway, before computers there was little alternative.

Many different forms of tables were introduced and are widely used today, but essentially they are of two types. Firstly, the problem of astronomical data is overcome by publishing ephemeris tables in the *Nautical Almanac*. From these it is possible to extract the predicted positions of sun, moon, navigational stars and planets for any second throughout the year. Secondly, the mathematics of working sights is reduced to simple addition or subtraction by sight reduction tables. Unfortunately the complexity of the mathematics has led to necessary compromises in the preparation of the tables, and ease of use has been traded against accuracy. This limitation of sight reduction tables has led to their appearance in several different forms, where the balance between accuracy and ease of use is differently struck. The tables referred to as UK publication AP3270 (US No. HO249) are a recent example, but these were developed originally for air navigation during the Second World War. Then they were intended to provide inexperienced airmen with a rapid means of finding their way about. However, they have since been modified for marine use and are now used routinely for yacht navigation.

THE LAST 40 YEARS

Forty years is a mere moment in the evolution of astro navigation, but one that has brought the most technological developments. One of the first applications of computers was in the preparation of tabular astronomical data and it is hard to imagine how the astronomical data available today could be supplied to a similar level of precision without their help. In this form at least, computer astro navigation has been around for many decades.

In the last decade computers have become pocket-sized and are now manufactured in large quantities for world markets at low unit prices. For astro navigation these computers present new possibilities. Sight reduction by Marcq Saint-Hilaire's method can be used to eliminate sight reduction tables and their compromises. This is explained in Chapter 5.

As an alternative to carrying around astronomical data on all objects, simple algorithms exist for predicting moon and star positions, and pocket computers provide an effective means of working them. These methods are described in Chapter 6, although it would be a mistake to believe that today's pocket computers could compete with the large number-crunching machines used by the *Nautical Almanac* office. However, the results should produce an error not greater than the equivalent of one nautical mile.

FOR THE FUTURE

One thing looks certain: pocket computers are going to have bigger memories and carry more functions. Whilst astro navigation may always be stuck with the vagaries of weather that occasionally blots out the sky, more powerful computers will make the system easier to use. A computer with an on-board clock and digital sextant interface would eliminate the need to type in time or sextant readings. More memory could be used to include programs with the facility to identify which stars or planets are being observed and give an indication of the accuracy of fixes obtained.

2. Tools and skills

Astro navigation without sextant, watch and some means of computing sights is certainly possible (see Chapter 9), but these are fundamental tools and I doubt whether any trans-ocean sailor would willingly put to sea without them. A small boat at sea is no place in which to learn that your equipment is inadequate, but the following guidelines may help in choosing what you take.

THE WATCH

There was once a time when timepieces for astro navigation meant clockwork marine chronometers costing a great deal of money. Cosseted away in lined boxes and requiring careful and consistent regular winding, they were still not always reliable. Today you could buy a suitable electronic watch from a street market stall for a tenth of the price. Such is progress.

Electronic watches are ideal for astro navigation. Even the simplest and cheapest can be quite useful and are available with an increasing number of features for little extra cost. Beware of having too many buttons to push – on a cold night, numb fingers will make operating them impossible, even if you can remember the correct sequence without having to refer to the manual. For astro work, a display with a date is useful, as is a stopwatch facility: you will need to record sight times to the nearest second. In this case make sure that it is possible to synchronize the stopwatch with the time normally shown by the watch. If you are taking a series of sights it is useful to be able to read off the time of each sight without having to resynchronize the stopwatch. Manufacturers selling their watches for timing races sometimes refer to this as a lap timer. My personal preference here is to keep the main watch protected below deck and, before taking sights, to synchronize it with a stopwatch for use on deck. The stopwatch has larger buttons and is no problem to operate even with gloves on.

If you do not have a stopwatch there are two alternatives. You can arrange for someone to help you take sights. One person is needed to make the observation and call to the other, who records the exact Greenwich Mean Time of the observation. The other alternative is to begin mentally counting seconds from the moment that you take the sight until you look at the watch and note down the time. To be reliable this method requires practice. Try counting the seconds as 101, 102, 103, 104 . . . instead of 1, 2, 3, 4 . . . and practise with a watch to see how consistent you can become.

Although watches may be sold as shock and water resistant,

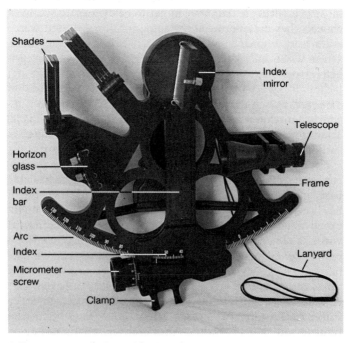

1 Sextant general view with named parts

it is no bad thing to treat them as though they were not. Shocks and temperature changes can disturb the timing, so it is a good idea not to wear the watch but to keep it protected below. Many watches that claim to resist water to depths of several hundred metres do not claim that any resistance is maintained if the buttons are pushed whilst it is immersed.

Whatever timepiece you choose to use it is unlikely to keep perfect time, but what is more important is that it gains or loses time by a reasonably consistent amount. Try to keep it at a fairly constant temperature and protect it from mechanical shocks. Check it regularly against radio time signals and keep a record of the amount by which it is losing or gaining. It is not necessary to correct the watch at every time signal; better to add or subtract the error when timing sights. The record should show how consistent the timing error is or if it gains or loses by irregular amounts.

Finally, although there are some solar powered watches, most need batteries so don't forget to carry a spare set. Even though they may not need replacing for years, if your watch still has its original batteries they could expire at any moment, as the watch may have been held in stock for many months before you bought it.

THE SEXTANT

The purpose of a sextant in astro navigation is to provide an accurate means of measuring the angle between an object in the sky and the horizon, though this is by no means its only application. It can also provide a useful navigation aid in coastal navigation where its accuracy in measuring vertical or horizontal angles far exceeds that of any other instrument likely to be to hand. Under reasonable conditions you might expect to be able to find angles with a hand bearing compass to within a few degrees. A sextant is likely to be at least 100 times better, but of course is unable to refer angles to magnetic north.

Traditionally sextants, like chronometers, were treated with fastidious care and a kind of reverence. They were only taken from their protective boxes before they were used and returned quickly afterwards. They were always picked up by the handle or frame, used with a neck lanyard, and never left lying on flat surfaces such as cockpit seats. Unlike the chronometer, recent technology has brought no fundamental design changes and the need for care in handling remains the same as ever. If a sextant is dropped, the degree scale (the arc) may become bent and accuracy is certain to be affected. Professional repairs are always expensive.

In recent years the most significant design change has involved the use of plastics. This has meant that instruments of an acceptable quality are now available at something like a quarter of the cost of their brass or aluminium framed counterparts. Plastic sextants are less prone to corrosion, and can be a very good buy, but the sextant is still the most expensive part of the astro navigator's gear.

Mechanical Sextant Adjustments

Even in the best sextants, small machining and alignment errors are always present. These can affect the accuracy of the instrument, but adjustments are provided to correct for the most serious of them. They are to some extent interdependent and should be adjusted in the order in which they are described below:

Perpendicularity error

This occurs when the index mirror is not quite at right angles to the frame. It is very easy to check when you know how to look at the sextant in the right way (see photograph 2). Hold the instrument horizontally with the index mirror uppermost and towards you. Look at the reflection of the arc in the index mirror. Now move the index arm until you can see part of the arc just beyond the right edge of the index mirror. This part should appear in the same line as that part indicated by the pencil in photograph 2 and reflected in the mirror. If it does not, then adjust the screw on the back of the mirror until it does.

Side error

Side error is another kind of perpendicularity error but this time it refers to the horizon glass. Set the index and micrometer tangent screw to zero and focus the telescope on a distant horizon. You should see the horizon as a continuous line cross-

2 Checking a sextant for perpendicularity error

ing both the silvered and clear parts of the index mirror. If the line has a step then adjust the micrometer screw until it is straightened out. Now gently rock the sextant from side to side so that you rotate it about the telescope axis. As you do this you should still see the horizon as a continuous line. If a step appears there is side error, and this is eliminated by adjusting the screw on the back of the index mirror that is furthest away from the frame.

An alternative method of checking for side error is to view a low star or the sun with the index bar just off zero. As you use the micrometer screw to bring the two images together, you should see them exactly coincide if there is no index error.

Index error
The effect of index error upon sextant readings is to produce an error, usually only a small number of minutes, that need to be added to or subtracted from each reading. If it is found to be much more than a few minutes, it should be corrected for by adjustment.

Whilst perpendicularity and side error may only need checking from time to time, index error needs checking every time that the sextant is used. It occurs if the horizon mirrors are not exactly parallel with each other when the index bar is set to zero. It is corrected in a similar way to the side error adjustment. Set both the index bar and micrometer screw to zero, then look again at the horizon. Index error is again shown by a break in the horizon line. This time it is corrected by adjusting the screw on the back of the index mirror that is closest to the frame.

To measure the index error, look at the horizon and adjust the micrometer screw until it forms a continuous line. The micrometer reading will now show the index error, but notice that it can be either off or on the arc. In other words, the error may be either positive or negative and will therefore need to be added to or subtracted from the sextant reading.

Index error adjustments may affect side error, but absolute precision is not essential. Better to accept a small amount of error and to work with it, adding or subtracting it from each reading as necessary.

Mathematical Adjustments

What is really needed is not the angle between the body and the horizon but the angle between the body and a true horizontal plane. By taking the horizon instead there are two factors to correct for. Firstly there is *dip*, which is the effect of the height above sea level from which the observation was made, and secondly *refraction*, caused by the bending of light as it passes into the earth's atmosphere. Fortunately they can be calculated and corrected by some straightforward mathematical expressions. These are given in Chapter 4, along with other corrections needed for sights of particular objects.

Difficulties Caused by the Weather

A small boat in a seaway is a rotten place for making precise sextant observations; larger vessels provide a more stable platform. Pitching and rolling make it difficult to hold the sextant steady. Choose a position near the centre of the boat and jam yourself against something substantial like a mast or deck house. You need to be safe and comfortable, so a kneeling or sitting position may help. The most effective improvements are made by practising and it is always surprising how much

3 Using a sextant

better you become after a few practice shots.

Mist makes the horizon indistinct and also leads to poor sight accuracy. If you have a choice, do not use anything other than a sharp and distinct horizon. However, if the clouds have blanketed the sky for days and you really need to know where you are, you will have to make the best of what you have. Do remember, though, to leave an appropriately large margin for error in your subsequent navigation.

THE COMPUTER

A computer is a device able to carry out a sequence of arithmetical and logical operations, which are specified by a list of instructions held within the computer memory.The distinction between pocket computers and calculators is not always clear. Calculators may carry out many scientific and other complex functions and could have a program memory for a few key strokes. In most cases memory capacity is small and the programming language will be specific to the particular machine. On the other hand, the computers that I am referring to here have very much larger memories and use a widely understood programming language. They can also be expected to contain facilities for listing programs and detecting and removing errors. Similar features are included in physically larger home or office personal computers, but in general their size, lack of environmental protection and power requirements make these unsuitable as computers for small craft. However, if you are sufficiently dedicated to computing afloat you will probably overcome these difficulties.

POCKET COMPUTERS FOR USE AT SEA

Pocket computers are truly pocket sized (not usually larger than 200 mm × 150 mm × 25 mm) and share a similar technology with watches. Included is a small liquid crystal display and an alphabetic and numerical key pad. Printers and cassette or disk drives can be added, and are useful in perfecting programs. However, they are certainly not essential and at sea could prove to be unreliable as it might be difficult to exclude damp from the mechanical parts.

Most pocket computers are not specifically intended for life at sea. Manufacturers usually make no special provisions for excluding water, and there is little doubt that the odd dunking would finish one off permanently. Fortunately the problem is easily overcome. Photograph 4 shows a case made from a block of teak with a neoprene sealing gasket. The case holds spare batteries and a stopwatch, and gives good protection from physical damage – it floats as well as keeping the water out. The same picture shows a plastic protective wallet made by Aquaman. This is claimed to give water resistance when submerged to 10 metres, not that this should be necessary, as the bag makes the computer float. A further advantage of using the wallet is that the computer can be operated through the plastic so it can be protected for the whole time it is at sea.

Pocket computer power requirements are modest, and battery lives are usually measured in years. But, as with digital watches, it is as well to carry spares. In order to economize on battery power, many computers are fitted with a facility that switches them off should no button be pressed after a delay of several minutes. When you wish to continue working, a reset button gets the program going again.

Although switching off the computer appears to shut it down completely, there is an internal connection that maintains a power supply to the computer memory. This is needed to maintain the program memory which would be erased if the supply were totally removed. This might happen when you replace the batteries, but manufacturers have different ways of overcoming the difficulty. Sometimes two or more separate batteries are fitted, and replacing them one at a time will still maintain the memory. Should the memory become erased

4 Pocket computers. Background left to right: the SHARP PC-1403; CASIO FX-720P enclosed together with a stopwatch in a home-made case made waterproof with a neoprene seal; CASIO FX-730P in a waterproof AQUASAC plastic bag. Foreground: SHARP PC-1246

for any reason, then keeping a copy of the listing will at least enable you to type it in again.

Typing in Program Instructions

There are two modes in which pocket computers can function. Set to 'run' the computer will carry out computations, but separate from this there is a 'write' or 'program' mode. In this state program instruction lists can be typed into the computer memory space. The computer must be able to accept programs in BASIC to run those listed here. Typing in instructions is a job needing careful attention to detail. Be careful not to confuse numbers 0 and 1 with the letters O and I. Syntax and punctuation are very important, and to help in finding mistakes computers can display error messages giving an indication of the nature and location of the problem. If you receive one of these messages, and, after having checked through your typing, you are convinced that you have made no such mistake, it is probable that you have entered a command in an unacceptable form. The manual supplied with the machine should provide a list of commands to show what alterations are needed.

A more insidious kind of mistake (fortunately rarer) is the sort that makes sense to the computer. Functioning may appear normal but wrong results are turned out. The trial result tables given later should help detect these problems. To find the location of the error try working through the computations a step at a time, checking each stage independently, perhaps in longhand or by pocket calculator. Most computers have at least some 'debugging' facilities that will ease this process.

As with watches, there is a tendency for manufacturers to cram the keyboard with as many buttons and functions as they can. For astro work, there is no necessity for a large number of specialized scientific function buttons, though it is important that those listed in Chapter 3 can be carried out within the program. This may require using the alphabetic keyboard to spell out functions like SIN, COS etc. instead of having a single button to do the job. For use with the programs in this book it is convenient if the computer will handle angles expressed in degrees.

STAR FINDERS

Of the thousands of visible stars only about fifty-seven are listed in the almanac for navigational use. In practice, most navigators will routinely use only a few of these and know where to find them almost instinctively. They will pick out these favourites by colour and brilliance as well as their relative positions in the sky. Learning to identify a few of these stars is easier with a plan of the night sky, the auxiliary star charts given in *Reed's Nautical Almanac*, or the Admiralty Star Finder and Identifier NP323. This consists of a pair of circular star plans, one of the northern and one of the southern hemisphere, and a set of transparent overlays. Having set up the device, the position of any star or planet can quickly be found from coordinates marked on the overlay.

3. Terms and symbols explained

An appreciation of the principles of astro navigation must begin with an understanding of the terms involved. Those listed below are explained in the context in which they are used later. In practice, astro navigators using computers need not be aware of several of the terms listed below, as they can be calculated within the program and are not printed out at the end.

First the system for defining positions on the surface of the earth.

LATITUDE AND LONGITUDE COORDINATES (LAT AND LNG)
Latitude and longitude are both angular distances measured at a point in the centre of the earth. Figs 3.1 and 3.2 show how these angles are used to construct a coordinate grid system covering the whole of the earth's surface, with lines of latitude running parallel with the equator. They are named as being north or south of the equator and given values ranging from zero at the equator to 90 degrees at the poles.

Longitude is named as east or west of the Greenwich meridian which, like all meridians, is a line joining the poles, though this one passes through the old Greenwich Observatory in London. Longitudes have values ranging from zero to 180 degrees, measured to the east or west of Greenwich.

THE NAUTICAL MILE
As Fig. 3.1 shows, latitude angles are measured at an imaginary point in the centre of the earth. The distance along the surface

of the earth made by one minute of angular measure at the centre is defined as one nautical mile.

Because the earth is not a perfect sphere but slightly squashed at the poles, the exact distance that one minute of latitude makes along the surface will be greater at the poles than it is at the equator. The difference is just 19 metres, but this can make quite a difference if you are comparing distances of several hundred miles.

Next we'll look at the system for defining the positions of objects in the sky. For navigational purposes it is convenient to think of the earth as being totally enclosed within a hollow 'sky sphere'. This sphere is at a very great distance from the

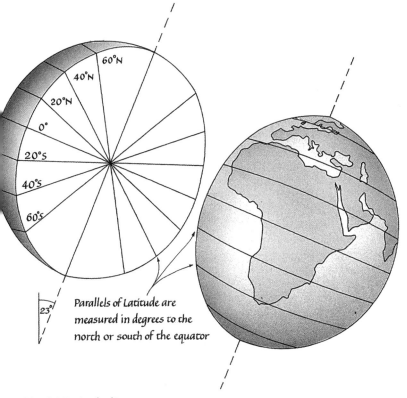

Parallels of Latitude are measured in degrees to the north or south of the equator

Fig. 3.1 Latitude diagram

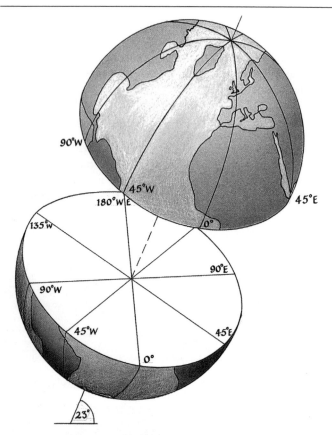

Longitude is measured in degrees to the east or west of the Greenwich meridian. This forms a line joining the north and south poles, and passing through Greenwich.

Fig. 3.2 Longitude diagram

earth, and the sun, moon, planets and stars all appear as points or blobs painted on the inside surface.

This is an oversimplification, of course. The moon in particular cannot be treated in this way, but for the moment this analogy does help show how the system for defining sky positions is simply an extension of the latitude and longitude system.

DECLINATION (DEC)

This is the sky equivalent of latitude. Simply imagine those rings of latitude projected out onto the sky sphere as in Fig. 3.3. Like earthly latitude, declination is also measured either to the north or south of a sky equator.

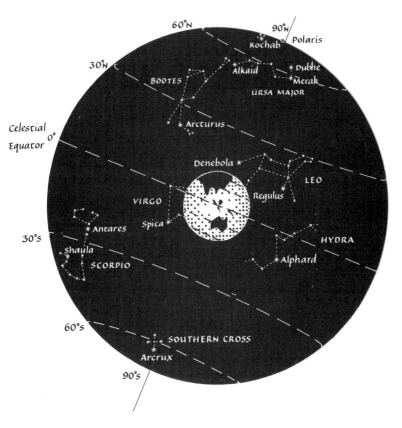

Like latitude, declination is measured in degrees to the north or south of the celestial equivalent of the equator. This diagram shows 0°, 30° and 60° latitude lines, along with some major stars in one half of the 'sky sphere'

Fig. 3.3 Declination coordinates

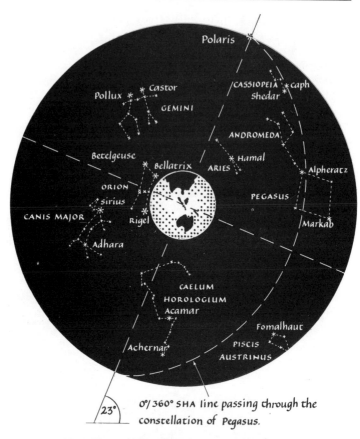

0°/360° SHA line passing through the
constellation of Pegasus.

SHA is comparable with longitude, but is measured only to the
west in degrees 0°-360° from the meridian passing through the
first point of Aries. Thus Markab has an approximate SHA of
14°, Pollux 244° and Hamal 328°.

Fig. 3.4 Sidereal Hour Angle

HOUR ANGLES
Hour angles are comparable with longitude but with a slight
difference. Whereas longitudes have values of up to 180
degrees and are measured either to the east or west of the
Greenwich meridian, hour angles are measured only to the

west and have values of up to 360 degrees. Three kinds of hour angle are important in astro navigation.

Sidereal Hour Angle (SHA)
In specifying the positions of stars it is convenient to refer their hour angles to a meridian line, the sky counterpart of the Greenwich meridian. As with all meridians, this line is in a plane passing through the poles. Instead of Greenwich, it passes through a point in the sky known as the First Point of Aries, or the vernal equinox. Actually the line no longer passes through the constellation of Aries, it goes through the square of Pegasus, but this doesn't matter because, like Greenwich, it is simply a reference point. See Fig. 3.4.

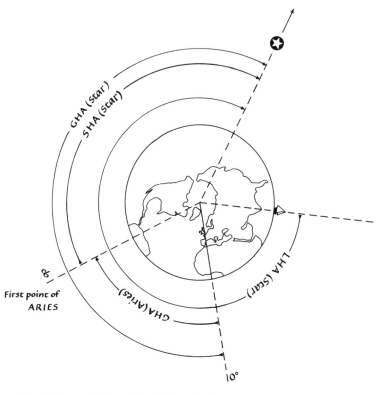

Fig. 3.5 Greenwich and Local Hour Angles

Greenwich Hour Angle (GHA)

Imagine the Greenwich meridian line projected out onto the sky sphere. Since the earth is rotating once in every 24 hours, in this time the line will pass every object in the sky once. At any moment in time the angular distance *westwards* from the meridian to any particular star is its Greenwich Hour Angle. Fig. 3.5 shows how the Greenwich Hour Angle of a star is the sum of its SHA and the GHA of Aries.

Local Hour Angle (LHA)

This is the angle between an observer's meridian and the meridian of an object in the sky. Again the angle is measured *westwards* from the observer's meridian. For stars the LHA is given by:

LHA (star) = GHA (Aries) + SHA ± observer's longitude

For the sun or moon the expression is:

LHA (body) = GHA (body) ± observer's longitude

In both of these expressions the convention is that easterly longitudes are added and westerly longitudes subtracted. If the LHA is greater than 360 degrees then 360 degrees is subtracted from it. See Fig. 3.5.

GREENWICH MEAN TIME (GMT)

Greenwich Mean Time is the time kept at any point along the Greenwich meridian. Here the middle of the day occurs around 1200 hours GMT and the middle of the night at about 2400 hours. Unfortunately the rest of the world does not see things in quite the same way. Midday for anywhere else in the world will be at a time difference from GMT which depends upon its longitude. North Island, New Zealand, for example, will see midday at around 2400 GMT. For local convenience different parts of the world adopt different time systems to compensate for this. For astro navigators this can produce all manner of difficulties unless they stick rigidly to Greenwich Mean Time, the only time system that will be used in this book.

You may encounter Universal Time (UT) which is synonymous with GMT or Universal Coordinated (UTC) Time upon which radio time systems are based. UTC may differ from GMT by 0.9 second which, for our purposes here, is small enough to ignore. Radio time signals can be obtained from radio stations across the world and the *Admiralty List of Radio Signals Volume 5* gives details of time systems and their transmission location, format and frequency.

Greenwich Date
This is the date kept at Greenwich and, as with GMT, the only date for astro navigators to be concerned with.

USING THE COMPUTER
In practice, computing sights is largely a question of applying a set of logical and mathematical manipulations to the above

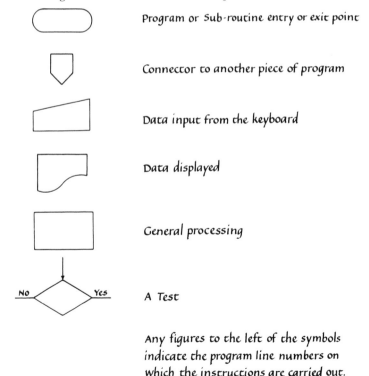

Program or Sub-routine entry or exit point

Connector to another piece of program

Data input from the keyboard

Data displayed

General processing

A Test

Any figures to the left of the symbols indicate the program line numbers on which the instructions are carried out.

Fig. 3.6 Flow diagram symbols

quantities. However, at first sight a computer program for doing the job can look daunting, especially if you are not used to working in BASIC. Here, flow diagrams can be a useful aid to understanding. The seven symbols shown in Fig. 3.6 will be used to help show how the input information is manipulated within the computer program.

Figure 3.7 is an example of how these symbols could be used to show the procedure for finding the LHA of a star, given SHA (star) and GHA (Aries).

Flow diagrams are no longer thought to be indispensable aids to program writing and not all programmers use them.

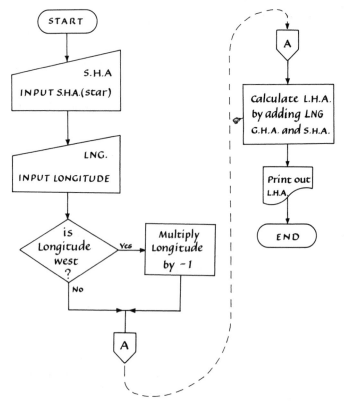

Fig. 3.7 Flow diagram example – calculating LHA from SHA and longitude

However, they do provide a way of breaking down programs into a simple sequence of events, and it is for this reason that they are included here. Those included are intended to show the underlying structure of programs, rather than all the fine mathematical detail. This can be obtained by examining the program listings directly.

Because there are several slightly different forms of BASIC in common use, the following explanation of the terms used in this book should help to clarify the intentions behind their use. Only the simplest BASIC instructions have been used here and these are common to very many versions. If differences exist with your particular machine then compare the use of the instructions given below with that in the computer handbook. It is most likely that there is a comparable instruction that will do the same job. First, though, a few words about the way in which variables are used and program lines numbered in the context of a pocket computer.

Variables
These are the labels or names used by the computer to identify items of data. Although most larger computers are able to accept several letters or whole words for variable names, many pocket computers will not, and for this reason only single, upper-case letters are used here. Listed below are those variables used to denote the coordinates described so far.

Latitude (LAT)	C
Longitude (LNG)	D
Declination (DEC)	E
Hour angles (SHA, GHA or LHA)	F

Appendix 3 gives a list showing how other letters have been allocated.

BASIC variables are subdivided into two types depending upon whether they are to hold numbers or groups of characters. *Numeric* variables appear as single letters but those for holding characters, known as *string* variables, are distinguished by adding a $ suffix. For example:

A = 45   would cause A to be loaded with the number 45.

A$ = "CARROTS" causes the string variable A$ to be
                loaded with the word CARROTS. Notice
                that it is necessary to place double
                quotation marks on either side of the
                word to identify it as a character string.

String variables can only be loaded with items enclosed between quotation marks. So a statement such as A$ = 45 would produce an error, as would an attempt to load a numeric variable with letters, eg A = "CABBAGE".

Line Numbers

Programs listed here have line numbers increasing in steps of 10. This follows the usual convention and allows you to introduce new lines should you wish.

Each of the following chapters is devoted to describing a separate part of the process of working sights to give position lines or fixes. At the end of each chapter there is a short segment of BASIC designed to carry out the steps outlined in the chapter. Do not worry if at first sight the choice of line number from which to start each segment appears to be fairly arbitrary. Each segment is not intended as a complete program in its own right but as a *subroutine* – part of a larger program connecting them together. It is the GOTO and GOSUBS that join them, and not a consecutive sequence of line numbers. This gives much greater flexibility, allowing you to make use of the subroutines for different purposes, and to construct programs to suit your own needs. However, Chapter 6 does give a connector or driver program.

Some smaller pocket computers cannot handle line numbers greater than 1000. In these cases, dividing line numbers by 10 will bring most programs within range. If you do alter line numbers then don't forget to alter, too, the program segments that refer to line numbers (ie GOTO or GOSUB – see below).

Working with Angles

Most pocket computers work easily with angles expressed in degrees. For this reason, and because navigators are well used to working in degrees, all program calculations assume this mode. Fractional angles, however, must be expressed in decimals of a degree and not as minutes, but this conversion is made within the program so the user is unaware of it taking place.

Some personal computers work in radians instead of degrees (360 degrees = 2 × Pi radians). If you wish to use one of these machines, see the notes in Appendix 1.

INSTRUCTIONS IN BASIC

Instruction	Meaning
+	Add.
−	Subtract.
*	Multiply.
/	Divide.
SQR	Square root.
↑	Raise to the power.
=	Equal to.
<>	Not equal to.
>	Greater than.
<	Less than.
>=	Greater than or equal to.
<=	Less than or equal to.
SIN	Returns the sine of an angle (ie an angle in degrees).
COS	Returns the cosine of an angle (ie an angle in degrees).
TAN	Returns the tangent of an angle (ie an angle in degrees).
ASN	Returns the inverse sine of an angle in degrees.
ACS	Returns the inverse cosine of an angle in degrees.
ATN	Returns the inverse tangent of an angle in degrees.
ABS	Returns the absolute (or modulus) of a number. ABS (2.68) = 2.68 ABS (−2.68) = 2.68
SGN	Returns with +1, 0 or −1

	depending on whether the number it is applied to is positive, zero or negative.
INT	Returns the integer part of a decimal fraction, eg INT (2.5) = 2; or for negative numbers INT (−2.5) = −3.
GOTO (a line number)	Makes program execution jump to another line number.
GOSUB (a line number)	Makes program execution jump to another line number. Execution then continues until a RETURN statement is encountered. Then a jump is made back to the point after which the GOSUB appeared.
ON (A) GOSUB . . .	This command causes program execution to jump to a subroutine that starts at a line number following the GOSUB. Several line numbers may follow the GOSUB but each must be separated with a comma. The subroutine that is executed depends upon the value of A. If A = 1 the first is executed. If A = 2 the second subroutine is executed etc.
ON (A) GOTO . . .	Behaviour is similar to the above except that the jump is to a line number and does not return after a subroutine.
IF (a condition) THEN . . .	If the bracketed condition is true the statement following THEN is executed.
END	End of program.

INPUT

This is a statement used to tell the computer to expect data to be typed into the keyboard. The data could be numerical or a string of alphabetic characters. INPUT A, for example, would cause program execution to wait whilst some numerical data was typed in. This would then be loaded into variable A. INPUT "HOW MANY?", A has a similar effect but this time HOW MANY? is displayed as a prompt by the computer whilst it is waiting. A comma or sometimes a semicolon is needed after the character string, as shown.

PRINT
Print statements cause the computer to display the data that is specified after the statement. For example, if A = 12, B = 6 and C$ = "TURNIPS" then:

PRINT A	would display	12
PRINT A/B	would display	2
PRINT C$	would display	TURNIPS
PRINT A/B; " CARROTS"	would display	2 CARROTS

Notice that in the above examples quotation marks are used to surround character strings and that several print statements can be included on one line provided that they are separated by a semicolon.

FOR . . . NEXT
This command creates a program loop that repeats all instructions appearing between the two words. The number of repeats is specified in a statement immediately following FOR. An example is the best aid to understanding this instruction. One is given at the end of Chapter 6.

REM
When the computer encounters a REM statement all characters that occur after it and in the same line are ignored. REM (or REMark) statements are usually used to include helpful comments in the program text which can be useful in understanding how a program works. Unfortunately they use up valuable memory space, which in pocket computers is often in short supply, so their use in programs here has been restricted.

LET

In most versions of BASIC the use of LET is optional and may be omitted. It is used to assign values to variables, for example LET A = 657.56 or LET H$ = "ASTRO". Some machines require LET to be used when variables are assigned following an IF . . . THEN statement. This is the only occasion on which it is used here, but you may need to check your computer manual to see if this is essential or if it has to be used in other circumstances.

Appendix 1 gives some additional commands that may be available on specific computers.

MEMORY SPACE

It is likely that future developments will bring pocket computers with much larger program memory capacity. 8K is typical of current machines, but older models may have only 1 or 2K (about an A4 page of program listing). In these cases you may well run short of memory before you have entered all of the listing that you want. If the deficiency is not too great there are several tactics that may help.

1 Look in the manual to see if there are any single complex instructions available that could incorporate several of the simpler instructions used in the listings.
2 Check in the computer manual to see how much memory is used up by the line number. You might save space by renumbering the lines to increase in steps of 1 instead of 10.
3 As a last resort you could try reducing the number of program lines by joining several consecutive lines together on a single line, but separated by a colon (:). This makes the program very difficult to follow but may make the difference between getting it to fit or not.

4. Taking sextant angles

Most astro sights involve measuring sextant angles, and this chapter shows how a single sight of an object produces a position line. It continues by using flow diagrams to show how the mathematical corrections mentioned in Chapter 2 can be built into a computer program, and it ends with the BASIC listings for a piece of program able to carry them out.

VERTICAL SEXTANT ANGLES USED IN COASTAL NAVIGATION
Tall coastal features (hilltops, towers etc.) are often marked on charts together with their heights above sea level. A measurement of the angle between sea level and the top of the object can be used to calculate how far away from it you are. Dividing the tangent of the angle by the height of the object gives the distance off (see Fig. 4.1). In using this method do not forget to adjust the object height to allow for the height of the tide and height above sea level from which the observation was taken. Also notice that the charted heights given for lighthouses are taken to the lantern and not to the top of the structure.

SEXTANT ANGLES IN ASTRO NAVIGATION
These terrestrial sights have much in common with astro sights which could be thought of as an extension of the technique to include objects of infinitely great height. A single sight also produces a position circle centred about the geographical position of the object. Figure 4.2 shows how the distance between an observer's zenith and an object in the sky is related to the distance between the observer and the geographical position of the object.

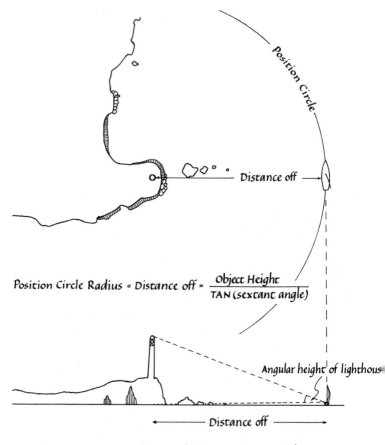

Position Circle Radius = Distance off = $\dfrac{Object\ Height}{TAN\ (sextant\ angle)}$

A vessel situated anywhere on the position circle would measure the same sextant angle of the lighthouse.

Fig. 4.1 Distance off tall coastal features can be calculated from a vertical sextant angle

This is a very simple technique for plotting lines and we will return to it in Chapter 8. Unfortunately most sights are taken of objects far away from the zenith where the position circle radius could be several thousand miles, and plotting such large distances with any accuracy is not easy. Only a very small part of the circle would appear on the chart, but with such a large radius this can be drawn as a straight line.

The next chapter shows how computing the position line can overcome the problems of trying to plot it, but first let's

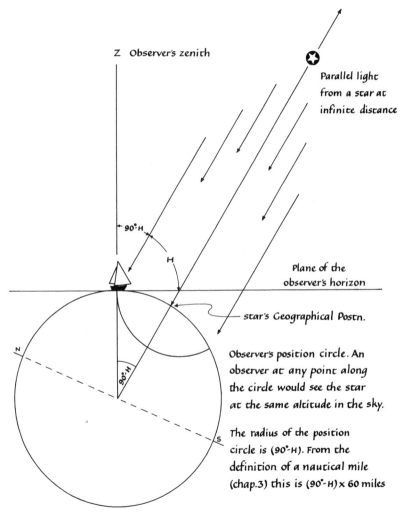

Fig. 4.2 Diagram showing how 90° altitude equals position circle radius

look at the corrections that need to be applied to all sights and other corrections that need only be applied to particular sorts of sight. To begin we need a short piece of program, which tells the computer which kind of sight it is expected to carry out.

THE MENU
The purpose of the menu is to give the variable, A, a value corresponding to the sight type.

Type of Sight	Value of A
Sun sight	1
Star sight	2
Moon sight	3
Planet sight	4

At various stages throughout the program (before the computer is about to carry out a step that differs for different sights) the value of A is checked and used to show what course to take.

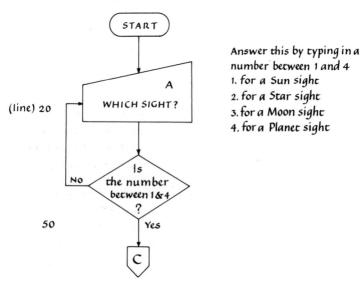

Answer this by typing in a number between 1 and 4
1. for a Sun sight
2. for a Star sight
3. for a Moon sight
4. for a Planet sight

Fig. 4.3 Menu – flow diagram

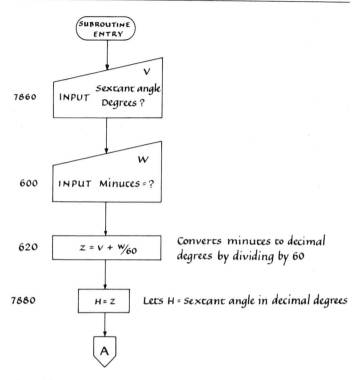

Fig. 4.4 Inputting a sextant angle and converting it to decimal form

Figure 4.3 shows the menu flow diagram. Although sufficient for our present purposes, it is a simplified description of the four menu program lines (20, 30, 40 and 50) at the end of the chapter. These contain a facility that allows a single letter to be typed in after the number denoting the sight type. The effect is to make A a negative number used in planning, a feature described on page 40.

COMPUTING WITH SEXTANT ANGLES

Pocket computers work most easily with angles expressed in degrees, but as navigators work mainly with degrees and minutes a short program segment is needed to carry out the conversion. The method chosen here involves typing in and entering the degrees part of the angle separately from the minutes part. This same format is used for entering angles later in the program, and enables degrees to be entered along with decimal fractions should this be necessary (Fig. 4.4).

INDEX ERROR

If the index error is off the arc (ie positive) it needs to be added to the sextant reading, and similarly if it is on the arc (negative) it is taken away. Figure 4.5 shows the procedure for entering the index error, which is first converted from minutes to decimals of a degree. Then the program requests a letter F or N to be typed in to show that it is either off or on the arc. Finally the corresponding addition or subtraction is carried out.

CORRECTIONS FOR DIP AND REFRACTION

Instead of the sextant angle of an object above the sea/sky horizon, what is really needed for working sights is the height of the object above a true horizontal plane. During World War II an instrument was perfected for aircraft navigation which achieved just this. The bubble sextant used a fluid containing an air bubble as its reference plane, rather like a spirit level. The idea sounds attractive and does not depend upon a visible horizon. Also these instruments are often available second-hand, so it is a pity that the motion aboard small boats makes their accuracy unacceptable.

Using the horizon leads to much more consistently accurate results but account must be taken of the effect of the height of the observer above sea level (dip) and of the refraction of light through the atmosphere. Refraction is greatest for objects on the horizon and zero for those immediately overhead. Both effects tend to increase the observed angles and so the corrections must be subtracted. Dip can be calculated from the following expression:

Dip correction = $0.0293 \times \sqrt{U}$
      where U = height of eye above sea level in metres

There are several formulae for calculating refraction. Some are accurate only for a restricted range of altitudes. But the following one, given in *Compact Data for Navigation and Astronomy* and attributed to G. G. Benett, is good for all angles.

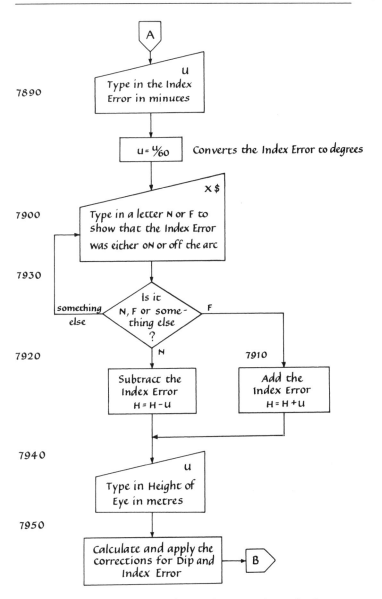

Fig. 4.5 Flow diagram showing how index error, dip and refraction corrections are applied to the sextant angle

Refraction correction = 0.0167/TAN (H + 7.31/(H + 4.4))
      where H = sextant angle

Refraction is dependent upon temperature and barometric pressure. This expression is valid for the standard conditions of 10°C and 1010 millibars pressure, although for practical purposes it is sufficient for most conditions likely to be encountered.* Should the conditions be vastly different then it would be as well to remember that refraction effects are less for objects higher in the sky and so those below, say, 25 degrees in particular should be avoided.

Corrections for both dip and refraction are applied in just one line of BASIC (line 7920) so finally making H the corrected altitude.

For sights of stars, and planets other than Mars or Venus, the sextant altitude corrected for refraction and dip is all that is required. However, there are additional corrections for sights of the sun, moon, Mars and Venus.

THE PLANETS

Of the four planets used for navigation, Saturn and Jupiter are both sufficiently far away from the earth to be treated in the same way as the stars. No additional corrections are needed. Mars and Venus, though, are rather closer, and the approximations made in considering them to be an infinite distance from the earth begin to break down. However, the additional corrections are still quite small. That for Mars is usually small enough to be neglected for most practical purposes, and that for Venus is also often ignored, but can be larger and up to 1 minute of arc or 1 nautical mile.

The actual values of these corrections vary with the time of year and the altitude of the sight, but figures may be obtained from the Altitude Correction Tables of the *Nautical Almanac*.

---

*Refraction effects under different conditions can be estimated by multiplying the correction given above by the factor f obtained from the following formula:

$$f = 0.28 \times P/(T + 273)$$

   where T = temperature in °C
      and P = pressure in millibars

Alternatively, *Compact Data for Navigation* gives some simple expressions from which they may be calculated.

To provide the opportunity to add in these small amounts, line 7990 in the sextant routine gives a prompt asking for any other corrections. At this point you can type in the required number of minutes.

THE SUN AND MOON

Both the sun and moon appear in the sky as discs. Although the position calculations require sun or moon altitudes to be measured to the centre of the disc, it is more accurate to use the sextant to measure the altitude to the upper or lower limb (edge) of the disc (Fig. 4.6). During some phases of the moon when the centre is missing it may be impossible to do anything else, though usual practice is to use the upper limb whenever pos-

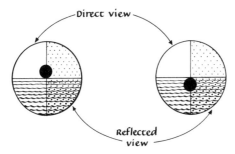

Upper and Lower Limb observations of the sun. Use shades to avoid ever looking directly at the sun.

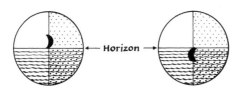

Upper and Lower Limb observations of the moon.

Fig. 4.6 Sextant telescope views of the Sun and Moon

sible. Correcting the altitude for the centre requires adding or subtracting the semi-diameter (radius).

Unfortunately the distance between the earth and the moon changes and this leads to considerable variation in the moon's semi-diameter. One solution is to look it up in the daily pages of a nautical almanac (see parallax corrections for the moon, below). You could also use the same approach for the sun, but here the variation is much less. Line 8000 sets the sun's semi-diameter at 16 minutes, and this simplification may lead to errors of 0.3 minute or 0.3 nautical mile. Errors would be greatest in December, January and mid-May to the end of August.

Semi-diameter addition or subtraction is a job carried out in lines 8030 and 8040.

Parallax Corrections for the Moon
If the errors made in assuming Mars and Venus to be of infinite distance from the earth are small, the same is certainly not true of the moon. No longer can it be thought of as a spot upon the inside of a sphere of infinite distance from the earth. In terms of astronomical distances it is only a short step away.

One effect of this is that the moon can appear to be in a different part of the sky when observed at the same time but from different positions on the earth. When taking sights of the moon this parallax effect needs to be taken into consideration. It is greatest when the moon is closest to the horizon and least when it is overhead. The correction is made in line 8010 and given by:

Parallax correction (in degrees) = S × 0.0612 × COS H
where H = the apparent altitude (in degrees)
          S = moon's semi-diameter (in minutes)

The daily ephemeris pages of nautical almanacs usually list the moon's semi-diameter (or the horizontal parallax, to which it is related) as follows:

Semi-diameter = 0.2724 × horizontal parallax

Figure 4.7 is a flow diagram showing the corrections for sun, moon and planet sights and completing the sextant subroutine.

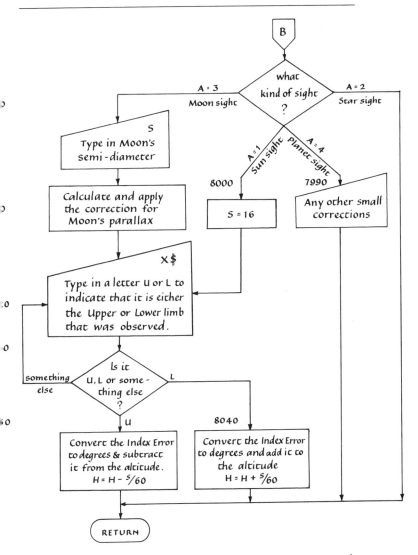

The completed sextant routine returns with H as the Sextant Angle corrected for Index Error, Dip, Refraction, Parallax and Semi-diameter.

Fig. 4.7 Flow diagram showing corrections for Sun, Moon and planet sights

OVERALL ACCURACY

The biggest and most likely source of error is caused by the boat's motion at sea, though it would be wrong to use this as an excuse for bad results. Under excellent conditions, an experienced operator could expect to get 95 per cent of his sights within 1 nautical mile, but this precision would be reduced as conditions deteriorate. Overall accuracy can be improved by taking several consecutive sights of the same object. This procedure is explained further on p. 78.

There are many correction factors that could be applied to different sights, but only the most important have been included here. Should you wish to program for others that make smaller contributions then *The Calculator Afloat* and the *Compact Data Book* provide details and formulae.

BASIC LISTINGS

(1) The Menu

```
20    INPUT "WHICH SIGHT ", X$: A=INT VAL(X$)
30    IF A<1 THEN GOTO 20
40    IF A>4 THEN GOTO 20
50    IF LEN (X$) >1 THEN LET A=-A
```

Variables required to be set on entry: None

Variables affected: A and X$

Variables set on exit:

A assumes a value of:
1 for sun sights
2 for star sights
3 for moon sights
4 for planet sights

(2) Sextant Subroutine

```
7850    Z=0
7860    INPUT "SEX.ALT DEG= ",V
7870    GOSUB 600
7880    H=Z
7890    INPUT "INDX.ER. ",U: U=U/60
7900    INPUT "N/F",X$
7910    IF X$="F" THEN LET H=H+U: GOTO 7940
7920    IF X$="N" THEN LET H=H-U: GOTO 7940
7930    GOTO 7900
7940    INPUT "HT.OF EYE ",U
7950    H=H-.0293*SQR U: V=H-.0167/ TAN (H+7.31/(H+4.4))
```

```
7970   ON A GOTO 8000,7980,8010,7990
7980   H=V: RETURN
7990   INPUT "OTR.COR= ",U: H=V+U/60: RETURN
8000   S=16: GOTO 8020
8010   INPUT "S.D.= ",S: V=V+S* 0.0612* COS H
8020   INPUT "U/L",X$
8030   IF X$="U" THEN LET H=V-S/60: RETURN
8040   IF X$="L" THEN LET H=V+S/60: RETURN
8050   GOTO 8020
```

Other subroutines called:	Conversion of degrees and minutes to decimal form subroutine (3) starting on line 600 (shown below)
Variables required to be set on entry:	A is given a value from 1 to 4 depending upon the sight type (see Menu 1 above).
Variables affected:	H, S, U, V, W, X, X$ and Z.
Variables set on exit:	H = sextant altitude corrected for index error, dip, refraction, semi-diameter and any other corrections.

(3) Conversion of Degrees, Minutes and Seconds to Decimal Form

```
600   INPUT "MIN=",W: IF Z=0 THEN LET X=0: GOTO 620
610   INPUT "SEC=",X
620   Z=V+W/60+X/3600:RETURN
```

Other subroutines called:	None.
Variables required to be set on entry:	U and Z. U is to be loaded with the number of degrees or hours. If Z is set to zero on entry, the request to input seconds is omitted.
Variables affected:	W, X and Z.
Variables set on exit:	Z = decimalized degrees (or hours), minutes and seconds.

5. Sight reduction

PREDICTING POSITIONS OF OBJECTS IN THE SKY

I don't suppose that Captain Marcq Saint-Hilaire had pocket computers in mind when he developed his method for reducing sights in 1875. Since then much effort has been directed at producing efficient tables for applying his technique, and most of today's systems use it in some shape or form. However, sight reduction is wonderfully suited to solution by pocket computer and can be done in just three program lines (7800, 7810 and 7820 at the end of the chapter).

Sight reduction provides the mathematical link between the positions of earthbound observers and positions of objects in the sky. If you know your LAT and LNG together with the DEC and GHA of some object in the sky you can predict the true bearing and height of the object above the horizon.

Several alternative formulae exist but the two chosen here are as follows. Notice that GHA and LNG are combined, as described in Chapter 3, to give LHA.

(1) Computed altitude =

[ASN (SIN LAT × SIN DEC) +

$\qquad$ (COS LAT × COS DEC × COS LHA)]

(2) Azimuth =

$$\text{ACS} \left[\frac{(\text{SIN DEC} - (\text{SIN computed altitude} \times \text{SIN LAT}))}{(\text{COS computed altitude} \times \text{COS LAT})} \right]$$

Should LHA be less than 180 degrees, then the azimuth is 360 minus that given in the above expression.

In using either of these expressions, the convention is that westerly longitudes as well as southerly latitudes and declinations are all entered as negative quantities.

Mathematically these formulae are providing the solution to a spherical triangle, often referred to as a PZX triangle and shown in Fig. 5.1. Here the observer's latitude is known, and so is the declination of the body. The angle between them, the LHA, is also known and these three pieces of information make it possible to solve the triangle and find any other angle or side.

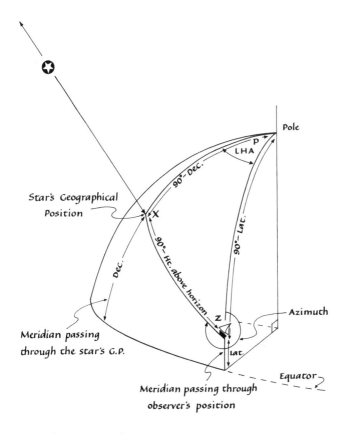

Fig. 5.1 The PZX triangle

THE CONNECTION BETWEEN PREDICTED AND OBSERVED ALTITUDES

At this stage you may be wondering what the point of this exercise could be. You are having to feed into the formulae the observer's LAT and LNG and if you knew that there would be no need to bother about taking sights.

Using the system in practice, the LAT and LNG that you feed in may be many miles away from the vessel's actual position. It is simply a convenient point somewhere on the chart that you are using and somewhere from which to begin plotting. The usual practice is to make this the vessel's dead reckoning position.

Since you are not expecting actually to be at the dead reckoning position it will come as no surprise when you find that the corrected sextant altitude differs from that predicted by the sight reduction formulae. This difference is known as the *intercept*. When expressed in minutes it represents the number of miles that your position line is from the dead reckoning position. Should the predicted altitude be greater than the corrected sextant altitude, the intercept is measured away from the observed body: if it is less, then it is measured towards it. Figure 5.2 shows these two possible cases.

PROGRAMMING FOR SIGHT REDUCTION

Looking through the listings at the end of the chapter you will see that there are rather more lines than just the three used to solve the PZX triangle. The remainder of the program is mainly concerned with loading LAT and LNG, DEC and GHA, guiding the program execution between the various subroutines and printing out the results.

Let's look first at subroutines (5) and (6) which are really four separate subroutines. Those starting on lines 500 and 800 are concerned with loading LAT, LNG, DEC and GHA into variables C, D, E and F respectively. However, LAT and DEC need to be made negative if they are southerly, as does LNG if it is westerly. This is done by typing in a single letter N, S, E or W in response to the question NAME? The two short subroutines beginning on lines 640 and 670 check that a valid character has been entered, and return immediately if no sign change is needed. If it is needed, the change is made by returning with

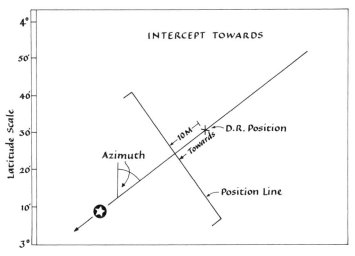

1. Draw a line passing through the D.R. posn. at the azimuth angle.

OTTING THE
•SITION LINE

2. Step off from the D.R. posn. along the azimuth, a distance
   equal to the intercept. This can be either towards (above),
   or away (below) from the body observed.

3. Draw in the posn. line at the intercept & right angles to the azm.

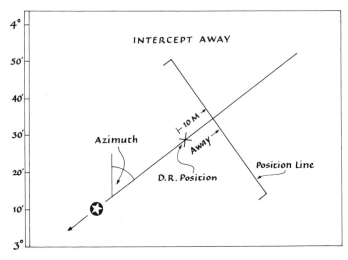

Fig. 5.2 Towards and away intercepts

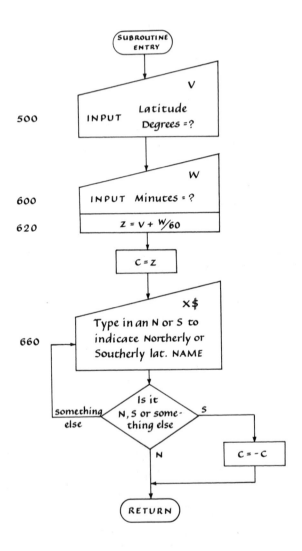

Fig. 5.3 Flow diagram showing how latitude is loaded into variable C. LNG, DEC, and GHA are loaded in a similar way into variables D, E and F

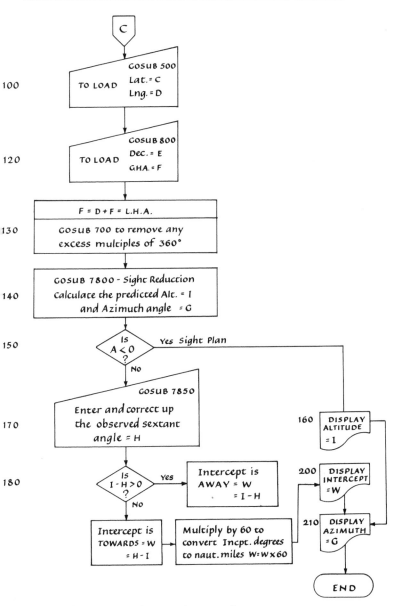

Fig. 5.4 Flow diagram showing the control program extension

$Z = -1$ and by multiplying this by the coordinate. Figure 5.3 shows the procedure for loading LAT, and the same procedure is used for loading the other coordinates.

Many computers are unable to carry out trig functions on angles with numeric values very much greater than 360°, which can occur when angles are added or multiplied. The short one-line subroutine in line 7000 subtracts unwanted multiples of 360°.

The remaining piece of program starting at line 100 is responsible for controlling access to the various subroutines and printing out the results. It follows directly on from the menu and its flow diagram is shown in Fig. 5.4. The subroutines encountered and the method for finding the intercept have already been described, but there are a couple of aspects worth explaining further.

Lines 150 and 160 are non-essential but do provide a facility useful in planning sights or in identifying stars or planets. It is very useful to be able to know at what bearing and altitude a star will appear, or to confirm its identity. In the menu, the question WHICH SIGHT? is answered with the sight type number and this may be followed by a single alphabetic character. In this case the sight variable A is given a negative value which lines 150 and 160 detect. It causes the sextant subroutine to be omitted and the program to end after printing out the predicted altitude and azimuth.

Lines 160, 200 and 210 are all concerned with displaying data and contain at least one INT statement. Here it is being used to limit the number of decimal places in the displayed data. There would be little point in displaying the azimuth to a millionth of a degree if you can only plot lines to within 1 degree or read your compass to within 5. This way of reducing the decimal places is quite cumbersome and it is probable that your particular computer will have a more appropriate command. However, such commands vary considerably between computer types and have therefore been avoided.

BASIC LISTINGS

(4) Control Program

```
100   GOSUB 500
120   GOSUB 800
130.  Z=D+F:GOSUB 7000: F=Z
140   GOSUB 7800
150   IF A> 0 THEN GOTO 170
160   PRINT "ALT = "; INT I; "DEG. ";INT((I-INTI)*600)/10: GOTO 210
170   GOSUB 7850
180   IF I-H> 0 THEN LET W=I-H: X$=" AWAY": V=-1: GOTO 200
190   W=H-I:X$=" T.WDS.": V=1
200   PRINT "INCPT = "; INT(W*600)/10; X$
210   PRINT "AZM = "; INT (G+.5): END
```

Subroutines called:	(5) LAT & LNG loading. (6) GHA & DEC loading. (7) Remove 360 degree multiples. (8) Sight Reduction
Variables required to be set on entry:	A - the menu variable.
Variables affected:	C.D,E,F,G,H,I,S,U,V,W,X,X$ and Z
Variables set on exit:	G = azimutn : W = intercept X$ is a towards or away indicator.

(5) Loading LAT and LNG

```
500   Z=0: INPUT "LAT   DEG= ",V: GOSUB 600: C=Z
510   Z=1: INPUT "LAT NAME?",X$: GOSUB 640: C=C*Z
520   Z=0: INPUT "LNG   DEG= ",V: GOSUB 600: D=Z
530   Z=1: INPUT "LNG NAME?",X$: GOSUB 670: D=D*Z
540   RETURN


640   IF X$="N" THEN RETURN
650   IF X$="S" THEN LET Z=-1: RETURN
660   INPUT "NAME?",X$: GOTO 640

670   IF X$="E" THEN RETURN
680   IF X$="W" THEN LET Z=-1: RETURN
690   INPUT "NAME?",X$: GOTO 670
```

Other subroutines called:	(3) Degrees and minutes to decimals on line 600.
Variables required to be set on entry:	None.
Variables affected:	C, D, U, W, X, X$ and Z.
Variables set on exit:	C = LAT; D = LNG.

(6) DEC and GHA Loading

```
800   Z=0:  INPUT "GHA  DEG= ",V:  GOSUB 600:  F=Z
810   Z=0:  INPUT "DEC  DEG= ",V:  GOSUB 600:  E=Z
820   Z=1:  INPUT "DEC NAME?",X$:  GOSUB 640:  E=E*Z
830   RETURN
```

Other subroutines called:

(3) Degrees and minutes to decimals beginning on line 600.
(5) Part only ie north or south naming routine beginning on line 640.

Variables required to be set on entry:

None.

Variables affected:

E, F, U, W, X, X$ and Z.

Variables set on exit:

E = DEC; F = GHA.

(7) Removing Excess Multiples of 360 Degrees

```
7000   Z=Z-(INT(Z/360)*360):  RETURN
```

Variables required to be set on entry:

Z = an angle with excess multiples of 360.

Variables affected:

Z.

Variables set on exit:

Z = the same angle less the excess multiples.

(8) Sight Reduction

```
7800   I=ASN(SIN C * SIN E + COS C * COS E * COS F)
7810   G=ACS((SIN E - SIN I * SIN C)/COS I /COS C)
7820   IF F<180 THEN LET G=360-G
7830   RETURN
```

Variables required to be set on entry:

C = LAT; E = DEC; F = LHA.

Variables affected:

I and G.

Variables set on exit:

I = computed altitude.
G = azimuth.

6. Adding some ephemeris

The program developed so far asks for DEC and GHA coordinates of the object on which the sight was taken to be typed in. These can be looked up, for the time of the sight, in the ephemeris pages of a nautical almanac. Not a difficult procedure, but it would be more convenient if the computer could be used to produce its own 'almanac' data. Computers were used to compile the almanac in the first place, so why not use one to do the same job at sea?

Even with the current rate of progress in computer design, it is still going to be quite a while before it is possible to replace the large machines used in preparing the nautical almanac with discount store pocket models. The accuracy with which almanacs are prepared is hard to reproduce, but algorithms for predicting the coordinates of navigational objects do exist and some, suitable for pocket computers, will give sufficient accuracy for small boat navigation.

This chapter gives two methods for adding ephemeris to the pocket computer programs. The first uses long-term algorithms for the stars, sun and moon, whilst the second method, published in *Compact Data for Navigation and Astronomy*, uses a set of tabular coefficients. These require monthly updating but the precision is greater and, because the methods are intended for calculators, the mathematics are shorter.

HANDLING TIME

In a program generating its own ephemeris, the date and time of the sight is used as a basis to compute DEC and GHA. The usual mixture of units (days, hours, minutes, months, seconds etc) for measuring time are not convenient to use in calculations. Much confusion can be avoided by sticking to a single unit and the day is the one used here. After typing in the date and time of the sight, the first job that the program must do is to convert these units to days and fractions of a day. This is the function of subroutine (9), Greenwich date and time.

Another detail that must be made clear is the point from which time measurements are to begin. Methods used in *Compact Data for Navigation and Astronomy* take it from 0 GMT on the 0th of the month in which the ephemeris is required. Put another way, this is midnight between the penultimate and last day of the previous month.

Some other ephemeris calculations cover a whole year and require time to be measured from 0 GMT on the 0th January in the required year. This is the approach used in the sun and Aries procedures described later.

Long-term algorithms often used by astronomers sometimes record events in Julian days, and these begin from noon GMT on 1st January in the year 4713 BC. Long enough ago, or so someone once thought, for no one to be much interested in anything that happened earlier. The moon algorithm is such an example, but to keep numbers of days to a more manageable size it has been adapted to accept days counted from 1980.

Subroutine (9) can cope with all three techniques, but before it can be used the current year must be loaded into variable N. Most pocket computers can store the values of variables after the power is switched off and, provided that N is used for no other purpose, this is a convenient way of avoiding having to enter the year every time the machine is used. Alternatively, if your machine does not have this facility, the year could be written into the program wherever N occurs, or included as a DATA statement.

Subroutine (9), Greenwich date and time, works as follows:

Line No.	Function
700	Loads the month and day into variables B and U.
710	Detects leap years.
720 to 740	Calculates the number of days from day 0 of the

current year, up to and including the current day.
750 Loads hours into V and uses the degrees and minutes subroutine to convert this into decimal hours. At the end of the line, B is left with the number of days and fractional days from the beginning of the year until the time of the sight, and Z with the number of days and fractional days from the beginning of the month until the time of the sight.

760 to 770 Calculates the number of days between day 0 in 1980 and the same day in the year of the sight (making due allowance for leap years). This is added to B to give L, the time variable needed for the moon algorithm.

CONTROL PROGRAM ADJUSTMENTS

In a program generating its own ephemeris, the control program needs to have a line added that will direct program executions to subroutine (9), Greenwich date and time. This could be as follows:

110 GOSUB 700

If you do not wish to use algorithms for computing the ephemeris on all objects (perhaps because your computer has insufficient memory or you would rather add them later) a conditional program jump is needed instead:

110 IF ABS (A) < 2.5 THEN GOSUB 700

This would have the effect of only loading Greenwich date and time for sun or star sights, that is, when A = 1 or = 2. Following on from this, line 120 causes the algorithm for the required body to be executed:

120 ON (ABS (A)) GOSUB 850, 900, 1000, 2000

Here, the value of A determines which of the four subroutines beginning on lines 850, 900, 1000 or 2000 is executed. These contain procedures for calculating the DEC and GHA of the sun, stars, moon or a planet, respectively. Should you wish to avoid the trouble of typing in some of the procedures given later in this chapter, then simply substitute the line number for 800, which is subroutine (6) for general purpose loading of DEC and GHA from an almanac data. For example:

120 ON (ABS(A)) GOSUB 850, 900, 800, 800

Lines 110 and 120 provide the flexibility to use whatever algorithms you choose for computing ephemeris on different objects.

STAR EPHEMERIS

Long-term methods for calculating star DECs and SHAs are given in both *The Calculator Afloat* and *Compact Data for Navigation and Astronomy*. However, the changes between one year and the next are very slight. As there are only some 57 stars that are generally used for navigation, it is usually convenient to keep a simple list updated annually and this is how this information appears in most nautical almanacs.

As was explained in Chapter 3, star GHAs are obtained by adding their SHAs to GHA Aries. Fortunately there is a very simple formula for calculating GHA (Aries):

$$GHA \text{ (Aries)} = R + (360.985647 \times B)$$

where B = number of whole days plus fractional days measured from day 0 of the year until the time of the sight. B is obtained from subroutine (9).

R = a constant for any given year and is given in Appendix 3.

This expression is written into line 930 – subroutine (10).

Subroutine (11) is another loading routine, concerned with loading star SHA into F. A call is made on the GHA (Aries) subroutine (10) which adds F to give GHA (star). Finally, at the end of line 910, a call is made to line 810 of subroutine (6) for loading DEC into variable E.

SUN EPHEMERIS

The sun is by far the most useful and most frequently used navigational body. Its availability is not restricted to the twilight times and no one ever confuses it with any other star.

The method used in subroutine (12) is based upon a short method described in *The Calculator Afloat* in which two annually updated quantities are required. These are the mean anomaly on day 0 of the year and the earth's longitude at peri-helion, but it is not essential to know how they are derived in order to use the method effectively. For an explanation see the two astronomy books mentioned in the bibliography. Appendix 2 lists their values in the form in which they are required in the sun ephemeris subroutine. Variables P and Q have been allocated with the exclusive job of holding these figures which, like N (the year), must be loaded before the program can be used.

MOON EPHEMERIS

The moon is the most complex navigational object to include in a pocket computer program. Nonetheless it is a useful object to have available. There is no problem in identifying it correctly and if it can be seen during the day during neap tides, a moon and sun position line will form a good angle of cut and so provide a reasonable fix.

The gravitational effects of both the sun and the earth upon the moon's rapid motion across the sky make it mathematically difficult to predict its position. To gain sufficient accuracy, the calculations involve adding many elements, each making an ever decreasing contribution. The long-term algorithm in subroutine (13) is based upon Nautical Almanac Office technical note No. 48 and is due to B. Emmerson. It has an inherent precision of about 1 minute and gives reasonable results for small boat navigation.

Unlike all other ephemeris calculations appearing here, the length and complexity of the moon algorithm does not lend itself to easy solution by calculator. However, a pocket computer is ideal for the job, as you only need to enter it correctly into the memory once to be sure of accurate future results.

EPHEMERIS FROM *COMPACT DATA FOR NAVIGATION AND ASTRONOMY*

This publication provides an alternative means of working sun and moon ephemeris as well as that for Venus, Mars, Jupiter and Saturn. The methods give a precision of about 0.1 minute and apart from the special case of the moon, the calculations for all objects follow a similar pattern. It would be quite feasible to construct an astro navigational program running

Table 1: SUN AND PLANETS, 1989

		January		February		March		April	
		GHA-GMT	DEC	GHA-GMT	DEC	GHA-GMT	DEC	GHA-GMT	DEC
		h	°	h	°	h	°	h	°
Sun	a_0	11.95123	-23.1004	11.77647	-17.4480	11.78940	-8.0555	11.92827	4.0608
	a_1	-0.25485	2.3290	-0.08123	8.9036	0.09794	12.0831	0.15996	12.4053
	a_2	0.03331	3.9950	0.11533	2.6764	0.07160	1.0384	-0.01444	-0.6242
	a_3	0.06347	-0.3110	0.00387	-0.7067	-0.02031	-0.6802	-0.02375	-0.5749
	a_4	-0.01912	-0.0798	-0.01173	0.0445	-0.00539	0.0621	0.00012	0.0394
check sum		11.77404	-17.1672	11.80271	-6.5302	11.93324	4.4479	12.05016	15.3064
Venus	a_0	13.61077	-21.9065	12.85339	-21.5142	12.32591	-12.5476	11.97119	2.3317
	a_1	-0.74673	-5.2783	-0.73689	6.0877	-0.46352	13.6203	-0.32308	16.0547
	a_2	-0.12506	5.5043	0.12418	5.5915	0.14762	2.9092	-0.01326	-0.3415
	a_3	0.09706	0.7421	0.05891	-0.8862	-0.03966	-1.1891	-0.06812	-1.1201
	a_4	-0.00555	-0.3801	-0.02939	-0.0808	-0.00928	0.0403	0.00896	0.0031
check sum		12.83049	-21.3185	12.27020	-10.8020	11.96107	2.8331	11.57569	16.9279
Mars	a_0	5.44376	8.1898	6.41857	14.6822	7.16005	19.7610	7.86945	23.5734
	a_1	1.11333	6.7676	0.91561	6.4026	0.78728	5.0524	0.68781	2.6905
	a_2	-0.13198	0.2008	-0.08908	-0.4842	-0.06861	-0.9909	-0.04178	-1.3929
	a_3	0.02717	-0.3062	0.01572	-0.2570	0.01357	-0.1977	0.01921	-0.1023
	a_4	-0.00518	0.0299	-0.00341	0.0330	-0.00139	0.0314	-0.00295	0.0300
check sum		6.44710	14.8819	7.25741	20.3766	7.89090	23.6562	8.53174	24.7987
Jupiter	a_0	2.99486	18.5631	5.06877	18.5783	6.75720	19.1941	8.46587	20.2708
	a_1	2.25471	-0.3717	2.02576	0.4047	1.84193	0.9542	1.69665	1.2033
		0.11082	0.2742	0.11989	0.3945	-0.09303	0.2360	-0.06061	0.0347

	h	o	h	o	h	o	h	o
						GHA-GMT	DEC	
				DEC		h	o	DEC
a_3	−0.01257	0.0511	0.01003	−0.0577	0.01385	−0.0826	0.01292	−0.0693
a_4	0.00580	−0.0256	0.00036	−0.0030	−0.00112	0.0067	−0.00168	0.0097
check sum	5.13197	18.5912	6.98603	19.3168	8.51883	20.3084	10.11315	21.4492
Saturn								
a_0	12.24708	−22.6117	14.02990	−22.4456	15.67978	−22.2538	17.58397	−22.0864
a_1	1.82983	0.1218	1.85778	0.2100	1.91831	0.2135	2.01717	0.1154
a_2	0.00297	0.0614	0.02562	0.0279	0.04358	−0.0218	0.05815	−0.0756
a_3	0.00877	−0.0092	0.00748	−0.0209	0.00626	−0.0245	0.00230	−0.0127
a_4	−0.00067	−0.0013	−0.00052	0.0011	−0.00087	0.0037	−0.00106	0.0042
check sum	14.08798	−22.4390	15.92026	−22.2275	17.64706	−22.0829	19.66053	−22.0551

Semi-diameter (S) of the Sun and HP of Venus and Mars

	o	o	o	o
Sun				
a_0	0.2716	0.2710	0.2695	0.2672
a_1	−0.0006	−0.0018	−0.0023	−0.0023
Venus HP				
a_0	0.0016	0.0015	0.0014	0.0014
a_1	−0.0001	−0.0001	0.0000	0.0000
Mars HP				
a_0	0.0025	0.0019	0.0016	0.0013
a_1	−0.0006	−0.0004	−0.0003	−0.0002

$$GHA^0 = 15((((a_4 x + a_3)x + a_2)x + a_1)x + a_0 + GMT^h)$$

$$DEC^0 = (((a_4 x + a_3)x + a_2)x + a_1)x + a_0$$

$$S^0, HP^0 = a_1 x + a_0 \qquad \text{where } x = (d + GMT^h/24)/32$$

Fig. 6.1 Sample data from *Compact Data for Navigation and Astronomy*

entirely on *Compact Data* methods but the example that follows is for planets only, though it does show how the methods can be programmed in BASIC.

Figure 6.1 shows a sample page of the data covering a four-month period. For each month there are five GHA and five DEC coefficients, referred to as A0 to A4, which need to be loaded into the computer. At the bottom of the column is a check sum that can be used to help show that they have been entered correctly. The example that follows uses an array to hold the data and a FOR and NEXT loop to load and retrieve it.

Different pocket computers use different procedures for handling arrays but, with some, care has to be taken to avoid clashing with other variables. In these cases an array of A with variables A(1), A(2) and A(3) will overwrite normal alphabetic variables B, C and D. The array chosen here uses Z, and because this falls at the end of the alphabet, the additional variables are non-alphabetic and hopefully should cause no problems. However, with some computers it may be necessary to 'declare' the array. Check the computer manual for details of any specific procedures.

For any one chosen planet the first piece of program required is a routine to aid loading its coefficients into these extra variables, which is given in subroutine (14a) beginning on line 9000. In (14a) line 9010 appears between a FOR and NEXT loop and is repeated 5 times. On each repeat U is incremented by 1. The effect of line 9010 is to print a request asking for the GHA coefficient A(U + 1) to be typed in. On leaving the loop, line 9030 prints the check sum. Lines 9040 to 9070 use an exactly similar process for loading the DEC coefficients.

USING THE COEFFICIENTS

Both DEC and $\dot{G}$HA – GMT are calculated from the coefficients using a polynomial expression of the form:

$$A0 + A1x + A2x^2 + A3x^3 + A4x^4$$

where x = $\dfrac{\left(\text{day of month} + \dfrac{\text{GMT hours}}{24} \right)}{32}$

Subroutine (9) returns with Z equal to the bracketed part of the above expression for x. This Z can be used in subroutine (14b) to return F with GHA and E with DEC. (14b) uses another FOR and NEXT loop to calculate and add each element of the polynomial. The term (24 * (Z − INTZ)) in line 2060 is rather cumbersome but extracts from Z the GMT in hours.

Line 9000 asks the question 'Coeff. OK?' The purpose is to provide a reminder that you need to have loaded the coefficients for the month and planet that you are observing. Answering the question with anything other than a letter Y will give the opportunity to load the correct coefficients by executing the loading routine (14a).

BASIC LISTINGS

Control Program Changes

```
110   IF ABS (A) < .....       see text for details. page 61
120   ON (ABS(A)) .....          "    "    "      "    page 62
```

(9) Greenwich Date and Time

```
700   INPUT "MONTH = ",B: INPUT "DAY = ",U
710   Y=0: IF INT(N/4)=N/4 THEN LET Y=1
720   IF B >2 THEN GOTO 740
730   B=INT((63-Y)*(B-1)/2): GOTO 750
740   B=INT(30.6*(B+1))-63+Y
750   Z=1: INPUT "GMT HRS = ",V: GOSUB 600: Z=U+Z/24: B=B+Z
760   L=INT(365.25*(N-1980))
770   L=L+1+B-Y: RETURN
```

Other subroutines called:	(3) Degrees (hours) and minutes to decimals beginning on line 600.
Variables required to be set on entry:	N = the current year.
Variables affected:	B, L, U, V, W, X, Y and Z.
Variables set on exit:	L = Days and fractional days from day 0 in 1980 until the time of the sight. B = Days and fractional days from day 0 of the year until the time of the sight. Z = Days and fractional days from day 0 of the month until the time of the sight.

(10) GHA Aries

```
930   Z=F+R+(360.985647*B): GOSUB 7000: RETURN
```

Variables required to be set on entry:	B = Days and fractional days from day 0 of the year until the time of the sight. F = Can be set to zero or to a star's SHA. R = The Aries coefficient (see text).
Variables affected:	Z
Variables set on exit:	Z = GHA (Aries) when F is set to 0 or Z = GHA (star) when F is set to SHA (star).

(11) Star DEC and SHA Loading

```
900   Z=0: INPUT "SHA   DEG= ",V: GOSUB 600: F=Z
910   GOSUB 930: F=Z: GOSUB 810: RETURN
```

Other subroutines called:	(3) Degrees and minutes to decimals beginning on line 600. (6) Part only ie DEC loading and naming beginning on line 810. (11) GHA Aries beginning on line 930. (7) Remove excess 360 degree multiples, beginning on line 7000.
Variables required to be set on entry:	B = Days and fractional days from day 0 of the year until the time of the sight.
Variables affected:	E, F, V, W, X, X\$, and Z.

Variables set on exit: E = DEC: F = GHA.

(12) Sun DEC and GHA

```
850   U=(.9856*B)+P
860   V=U+(1.916 * SIN U)+(0.02 * SIN (2*U))-Q
870   E=ASN(.3978 * SIN V): X=ATN(.9175 * TAN V)
880   IF SGN SIN X <> SGN SIN V THEN LET U=U+180
890   F=360*(B-INTB)+U-X-180-Q: RETURN
```

Variables required to be set
on entry: B = Days and fractional days
                                from day 0 of the year until
                                the time of the sight.
                                Z = GMT in hours and
                                decimal fractions of an hour.
                                P and Q annual update
                                coefficients (see text for
                                details).

Variables affected: E, F, U, V, X and Z.

Variables set on exit: E = DEC: F = GHA.

(13) Moon DEC and GHA

```
1000   L=(L+29218.5)/36525: L=L+L*.0000000222
1010   U=296.1+477198.849*L+.0092*L*L:   Z=U: GOSUB 7000: U=Z
1020   V=358.48+35999.05*L:               Z=V: GOSUB 7000: V=Z
1030   W=11.25+483202.025*L-.0032*L*L:    Z=W: GOSUB 7000: W=Z
1040   X=350.74+445267.114*L-.0014*L*L:   Z=X: GOSUB 7000: X=Z

1050   Y=270.434+481267.8831*L+ .004* SIN(193-132.9*L)
1060   Y=Y+6.289* SIN (U)-            1.274* SIN (U-2*X)
1070   Y=Y+ .658* SIN (2*X)+          .214* SIN (2*U)
1080   Y=Y- .186* SIN (V)-            .114* SIN (2*W)
1090   Y=Y- .059* SIN (2*U-2*X)-      .057* SIN (U+V-2*X)
1100   Y=Y+ .053* SIN (U+2*X)-        .046* SIN (V-2*X)
1110   Y=Y+ .041* SIN (U-V)-          .035* SIN (X)
1120   Y=Y- .03 * SIN (U+V)-          .015* SIN (2*W-2*X)
1130   Y=Y- .013* SIN (U+2*W)+.       .011* SIN (U-2*W)
1140   Y=Y- .011* SIN (U-4*X)+        .01 * SIN (3*U)
1150   Y=Y- .009* SIN (2*U-4*X)+      .008* SIN (U-V-2*X)
1160   Y=Y- .007* SIN (V+2*X)+        .005* SIN (U-X)
1170   Y=Y+ .005* SIN (V+X)+          .004* SIN (U-V+2*X)
1180   Y=Y+ .004* SIN (2*U+2*X)+      .004* SIN (4*X)
1190   Y=Y- .004* SIN (3*U-2*X)+      .003* SIN (2*U-V)
1200   Y=Y+ .003* SIN (U-2*W-2*X)-    .002* SIN (2*U+V-2*X)
1210   Z=Y- .002* SIN (U+X)-          .002* SIN (2*V-2*X)
1220   GOSUB 7000: Y=Z

1240   F=  5.128* SIN (W)+            .281* SIN (U+W)
1250   F=F- .278* SIN (-U+W)-         .173* SIN (W-2*X)
1260   F=F+ .055* SIN (-U+W+2*X)-     .046* SIN (U+W-2*X)
1270   F=F+ .033* SIN (W+2*X)+        .017* SIN (2*U+W)
1280   F=F- .009* SIN (-U+W-2*X)-     .009* SIN (-2*U+W)
```

```
1290   F=F-  .008* SIN (V+W-2*X)-      .004* SIN (2*U+W-2*X)
1300   F=F+  .004* SIN (U+W+2*X)+      .003* SIN (-V+W-2*X)
1310   F=F+  .002* SIN (-U-V+W+2*X)+.002* SIN (-V+W+2*X)
1320   Z=F-  .002* SIN (U+V+W-2*X)
1330   GOSUB 7000: F=Z

1350   E=    .95075+               .05182* COSRAD (U)
1360   E=E+  .00953* COSRAD (U-2*X)+   .00784* COSRAD (2*X)
1370   E=E+  .00282* COSRAD (2*U)+     .00086* COSRAD (U+2*X)
1380   E=E+  .00053* COSRAD (V-2*X)+   .0004 * COSRAD (U+V-2*X)
1390   E=E+  .00032* COSRAD (U-V)-     .00027* COSRAD (X)
1400   E=E-  .00026* COSRAD (U+V)-     .0002 * COSRAD (U-2*W)
1410   E=E+  .00017* COSRAD (3*U)+     .00017* COSRAD (U-4*X)
1420   E=E-  .00011* COSRAD (V)+       .0001 * COSRAD (2*U-4*X)
1430   E=E-  .00008* COSRAD (2*U-2*X)- .00008* COSRAD (V+2*X)
1440   E=E+  .00008* COSRAD (2*U+2*X)+ .00007* COSRAD (4*X)
1450   Z=E+  .00006* COSRAD (U-V+2*X)- .00006* COSRAD (U-V-2*X)
1460   GOSUB 7000: E=Z

1480   S=16.35* E: Z=259-1934.1* L: GOSUB 7000
1490   V=23.452-.013* L+.003* COS Z: U=Y-.005* SIN Z
1500   E=ASN ((COS F* SIN U* SIN V)+ SIN F* COS V)
1510   X=SIN U* COS V- TAN F* SIN V
1520   Y=COS U: Z=ATN (X/Y)
1530   IF Y < 0 THEN LET Z=Z+180
1540   GOSUB 7000: W=Z
1550   F=0: GOSUB 930: F=Z-W: RETURN
```

Other subroutines called:	(7) Remove excess 360 degree multiples, beginning on line 7000. (11) GHA (Aries), beginning on line 930.
Variables required to be set on entry:	B = Days and fractional days from day 0 of the year until the time of the sight. L = Days and fractional days from day 0 of 1980 until the time of the sight.
Variables affected:	E, F, S, U, V, W, X, Y and Z.
Variables set on exit:	S = Moon's semi-diameter in minutes. E = DEC, F = GHA.

NB

Line 8010 in the sextant subroutine (2) asks for the Moon's semi-diameter to be typed in. The Moon algorithm computes semi-diameter, so making this request unnecessary. Line 8010 should be shortened as follows:

```
8010   V=V+S* .0612* COSRAD H
```

(14a) Loading Routine for Compact Data Coefficients

```
9000   FOR U=1 TO 5
9010   PRINT "GHA COEFFICIENT  A";U-1;: INPUT Z(U)
9020   NEXT U
9030   PRINT "CHECK SUM ="; Z(1)+Z(2)+Z(3)+Z(4)+Z(5)

9040   FOR U=1 TO 5
9050   PRINT "DEC COEFFICIENT  A";U-1;: INPUT Z(U+5)
9060   NEXT U
9070   PRINT "CHECK SUM ="; Z(6)+Z(7)+Z(8)+Z(9)+Z(10)
9080   RETURN
```

Variables affected:	$Z(1)$ to $Z(10)$ and U
Variables set on exit:	$Z(1)$ to $Z(5)$ loaded with GHA–GMT coefficients $A(0)$ to $A(4)$ $Z(6)$ to $Z(10)$ loaded with DEC coefficients $A(0)$ to $A(4)$

(14b) Compact Data Method for Planet DEC and GHA

```
2000   INPUT "Coeff. OK  (Y)",X$
2010   IF X$<> "Y" THEN GOSUB 9000
2020   E=0: F=0: V=Z/32: FOR U=0 TO 4
2030   F=F+(Z(U+1)*V↑U)
2040   E=E+(Z(U+6)*V↑U)
2050   NEXT U
2060   F=15*(F+(24*(Z-INTZ)))
2070   RETURN
```

Variables required to be set on entry:	Z = Days and fractional days from day 0 of the month until the time of the sight.
Variables affected:	E, F, U, X$ and V.
Variables set on exit:	E = DEC: F = GHA

7. Latitude and longitude fixes

In Chapter 5 we saw how the azimuth and intercept obtained from an astro sight gives a position line, and Fig. 5.2. showed how it was plotted on a chart. Fixing the position of a vessel requires sights from at least two objects and they should preferably appear in the sky with a horizontal separation of between 60 and 90 degrees. This will correspond to the angle at which their position lines will intersect when plotted on the chart.

Plotting is the conventional way of obtaining fixes and the LAT/LNG coordinates of the position line intersection can be read off from the chart LAT and LNG scales. However, an alternative to the chart work is to calculate the fix. In the following formulae, away intercepts are entered as negative quantities and, following the usual convention, southerly latitudes and westerly longitudes are also negative.

Fix LAT = C−(((W × SIN J)−(K × SIN G))/60/SIN (G−J))

Fix LNG = D + (((W × COS J)−(K × COS G))/60/SIN (G−J)/ COS C)

where:

C and D = LAT/LNG coordinates of the dead reckoning position or other position from which the sight was worked.

> J = Azimuth of the first sight
> K = Intercept of the first sight
> G = Azimuth of the second sight
> W = Intercept of the second sight

This formula requires that both sights be made at the same

time and worked from the same position, but in practice this is unlikely. Nonetheless, this will not cause difficulties if the distance travelled between sights is small when compared with the anticipated accuracy of the fix.

Finding two suitably placed objects from which to obtain a fix is not too difficult at morning and evening twilights on clearish nights. There may even be a choice of several stars, the moon or the odd planet. During the day, the sun is the most obvious navigational object but then there is the problem of finding something else to cross it with. Occasionally the moon can be seen and on even rarer occasions Venus may also be visible. Here, it is helpful if you know exactly where to aim the sextant telescope, and for this the sight planning facility described in Chapter 5 could be useful.

Sometimes, though, circumstances make it impossible to take two suitable sights before the distance travelled by the boat becomes significant. Perhaps cloud cover prevents two suitable objects from appearing at the same time, or, in the case of high-speed craft, perhaps the distance moved between consecutive sights is too great. In these cases it may still be possible to make an accurate fix by the method of transferring the first position line.

TRANSFERRED POSITIONS

In times of heavy cloud cover, whole days may pass when you are lucky to see one useful object let alone two. It is on occasions such as these that you have to make the best of what is available, especially if you are expecting to make an imminent landfall or have some other urgent need to fix a position. Typically the sun might only appear through the clouds for a few brief moments, just enough for one sight. However, if two such opportunities occur, say one in the morning and another in the afternoon, a good fix is still possible. In this way the sun's azimuth will have changed sufficiently for the two position lines to intersect at an acceptable angle, though the boat's movements between sights have to be taken into account when plotting the fix. The method is to apply dead reckoning techniques to shift the first position line by the distance and in the direction that the vessel has moved between sights. This is shown in Fig. 7.1.

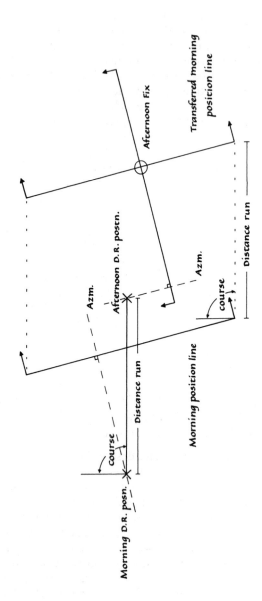

Fig. 7.1 Transferring a position line

LAT/LNG FIXES BY COMPUTER

The BASIC listings given at the end of this chapter extend the program developed earlier to enable fixes to be computed from paired sights, and with the option of including course and distance corrections between sights. The LAT/LNG fix formulae described above are both incorporated in subroutine (16) which appears at the end of this chapter, along with the changes and additions to the control program needed to implement it.

The earlier program functions as before but the effect of these additions is that after printing out an azimuth and intercept, the question POSITION CHANGE? appears. This can be answered in one of three ways.

(1) *By pressing the execute button without entering any letters.* This causes the computer to begin working another sight from the same dead reckoning position as the first.

(2) *By entering a single letter N (for no).* Again this causes the computer to begin working another sight, also from the same dead reckoning position. This time the second sight position line is crossed with that of the first and the program ends after printing out the fix LAT and LNG.

(3) *By entering a single letter Y (for yes).* This option also ends after printing out a two-sight fix, but this time there is an opportunity to allow for a change of position between sights. The two questions COURSE? and DISTANCE? are answered by typing in the true compass course and distance travelled between sights.

Figure 7.2 provides a flow chart description of these extra lines.

LIVING AND WORKING WITH ERRORS

Errors are an inherent part of any measurement. Whether it be taking sextant altitudes or weighing vegetables there will always be some uncertainty associated with the final result, but the significance of the error depends upon the purpose for which the measurement is required. A sextant angle error of half a degree corresponds to 30 miles; in mid-ocean this might not affect your navigation too much, but if you were attempt-

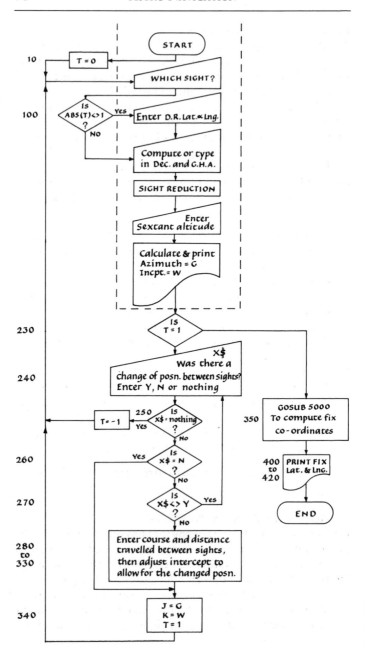

ing to reach a small island landfall it could make the difference between seeing it or going on past. Were this to happen your difficulties would be compounded if you had not considered the possibility of the error occurring. It is important to examine carefully all possible sources of error and their effects on the result.

In astro navigation the most significant errors are usually made in measuring sextant angles. Generally, these depend upon the experience of the navigator and the weather conditions prevailing at the time of the sight. Under ideal conditions 95% of sights made by an experienced operator could be expected to be within 1 mile; this uncertainty could be represented by making the position line a 2 mile wide band with the 95% chance that the vessel lies somewhere within it. Under worsening conditions its width will increase but it is important to know what results you personally can expect to achieve under differing conditions. This means making use of every opportunity for taking sights (even on occasions when the vessel's position is known) then examining your performance and striving to make improvements.

In general, errors can be grouped into three types. Firstly, there are the unexpected and sometimes quite large errors that may result from a miscalculation or an equipment fault. Whilst these cannot be prevented, they can be reduced by a methodical approach and occasional checks by alternative methods. With a little practice and experience, you can become aware of the range of results you might reasonably expect to occur, this being used to trigger an investigation should they not appear.

Secondly, there are systematic errors that always affect the results by a constant amount. Examples here are index error, calibration errors and the kind of personal errors that lead particular individuals to produce results that are always in error by a constant amount. Systematic errors can generally be corrected for once you are aware that they exist.

Finally, there are random errors, and in astro navigation by

Fig. 7.2 Obtaining LAT/LNG fixes from paired sights. The program developed in previous chapters is shown within the dotted box

far the most significant are those caused by motion of the boat while taking sights. In principle they are reduced if you are able to take several sights and average the results.

If conditions permit, it is always a good idea to take several sights of each object you intend to use. Though there is no sure way of saying which are the most accurate, the conventional technique is to plot the results on a piece of graph paper using sight time and sextant altitude as the two axes. This makes any bad sights more obvious and only those that fall close to a 'best fit' curve need be used. If you are using a pocket computer to work sights then the time spent on calculations is minimal and an alternative technique is to reduce *all* the sights you take. This will give a selection of intercepts, from which you will be able to pick out those that deviate most from the mean.

PLOTTING VERSUS CHARTWORK
A few words of caution. When using the LAT/LNG fix procedure it is sometimes easy to lose sight of the limitations of the data upon which the computer bases its results. For example, nothing prevents you obtaining a fix from a pair of stars whose azimuth angles are around 90 and 260 degrees. Their position lines will be nearly parallel and the fix accuracy totally useless. On the other hand, were the same results to be plotted on the chart the problem would be immediately apparent. Sketching the plot can provide a way of reducing these difficulties as well as giving a rough check on the computed results.

FIXES FROM MORE THAN TWO POSITION LINES
The program given here is capable of computing fixes from a pair of position lines, but better fix accuracy is obtained by taking sights of more than two objects. In these cases the best fix is derived from the 'cocked hat' shape formed from the position line intersections. Figure 7.3 shows examples of fixes obtained by plotting several position lines.

Plotting is a simple procedure and sights a long way away from all others, or those with small position line intersections, show up easily. Whilst it would be possible to arrange for the computer to do something similar, the program would become much more complex and involve statistical methods beyond the scope of this book. Here again conventional plotting tech-

niques are the simplest means of displaying the results so that
the limitations of the data upon which they are based can be
most easily appreciated.

BASIC LISTINGS

(15) Control Program Additions and Extension

```
  10   T=0
 100   IF ABS (T)<>1 THEN GOSUB 500
 210   PRINT "AZM = "; INT G: IF A<0 THEN GOTO 110


 220   IF V=-1 THEN LET Z=G+180: GOSUB 7000: G=Z
 230   IF T=1 THEN GOTO 350

 240   INPUT "POSITION CHANGE ? ",X$
 250   IF X$= "" THEN LET T=-1: GOTO 20
 260   IF X$= "N" THEN GOTO 340
 270   IF X$<>"Y" THEN GOTO 240

 280   INPUT "COURSE ",X: INPUT "DISTANCE ",V
 290   U=90+G
 300   IF SGN COS U=1 THEN LET Z=U+180: GOSUB 7000: U=Z
 310   U=V* SIN (U-X): IF SGN SIN G=-1 THEN LET U=U*-1
 320   W=W+U
 330   IF W<0 THEN LET W= ABS (W): Z=G+180: GOSUB 7000: G=Z

 340   J=G: K=W: T=1: GOTO 20

 350   GOSUB 5000
 360   X$=" S": IF SGN C>0 THEN LET X$=" N"
 370   Y$=" W": IF SGN D>0 THEN LET Y$=" E"
 380   C=ABS (C): D=ABS (D)

 400   PRINT "LAT = ";INT C;" DEG ";INT((C-INTC)*600)/10;X$
 410   PRINT "LNG = ";INT D; "DEG ";INT((D-INTD)*600)/10;Y$
 420   END
```

Subroutines called: (3) Degrees and minutes to
                             decimals.
                             (5) LAT and LNG loading.
                             (7) Remove excess 360
                             degree multiples.
                             (16) LAT and LNG fix.

(16) LAT/LNG Fix

```
5000   C=C-(((W* SIN J)-(K* SIN G))/60/SIN (G-J))
5010   D=D+(((W* COS J)-(K* COS G))/60/SIN (G-J)/COS C)
5020   RETURN
```

Variables required to be set
on entry: J = Azimuth of the first
                             sight.

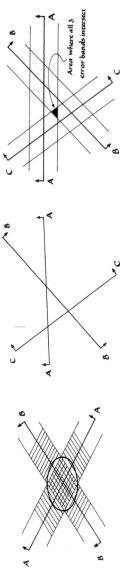

Position lines A & B are drawn together with their bands of 95% certainty. The fix is most likely to lie within the ellipse at their intersection though its exact location cannot be known.

3 or more position lines seldom intersect at the same point, but enclose a triangle or 'cocked hat', within which the fix is most likely to lie.

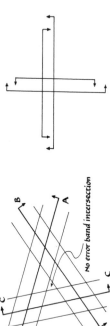

Drawing in the error bands shows the part of the triangle with the highest probability of containing the fix.

Drawing in the error bands can also help to show up mistakes. In this case there is no common high probability area where all the bands intersect.

Abnormal atmospheric refraction effects can lead to systematic errors. These can be reduced by taking sights from pairs of objects about 180° apart.

Fig. 7.3 Fixes from several position lines

K = Intercept of the first sight corrected for any change of position.
W = Intercept of the second sight.
G = Azimuth of the second sight.

Variables affected: C and D.

Variables set on exit: C = fix LAT, D = fix LNG.

8. Emergency astro navigation

Working with errors in practical navigation implies an acceptance that the procedures and equipment you are using are not always totally accurate. The next step to consider is how you would cope with the failure of one item, or maybe even all of your equipment. This doesn't only mean being pooped in the southern ocean or some rare or unlikely disaster. Minor accidents such as someone treading on your sextant, losing your watch overboard or even a small fire can have a similar effect upon navigation, but the impact will be minimized if you have considered and prepared for the possibility. A few plotting sheets, scientific calculator, cheap plastic sextant, watch, pencils, rubber, ruler, compass and relevant formulae could provide all that is needed if they are in a waterproof container and kept ready for use.

In an emergency, the equipment and procedures with which you are left may not reach the standards of convenience and accuracy to which you may have been accustomed. Nonetheless almost normal navigation may still be possible with quite a few bits missing, though you will need a certain amount of mental flexibility and a good understanding of the basic principles. In this chapter we will look at a few of the possibilities you might consider after the loss of some apparently 'essential' items of equipment.

LOSS OF THE COMPASS
With the computer and watch still intact true bearings of any object, day or night, can be determined by running the sight plan program option described at the end of Chapter 5. Should

you have a choice, use objects lower in the sky, as their bearings are easier to measure. The technique is also useful for compass checking either as a routine or should its accuracy be suspected, perhaps following a thunderstorm.

Without the computer you could use the star finder (NP 323) to do a similar job but with less accuracy. For setting it up GHA (Aries) is needed which can be found either by calculation or from the almanac. Once this is done it can be used to find altitudes and azimuths of any object in the sky.

A traditional way of compass checking is to use the rising and setting angle of the sun. For this you need an idea of your latitude and the sun's DEC. This can be used in the following formula to calculate the sun's amplitude, which is its bearing measured from true east or west.

$$\text{AMPLITUDE} = \text{ASN (SIN DEC / COS LAT)}$$

When using this formula if DEC is northerly then the amplitude is measured to the north of east or west, but measured to the south for southerly DEC. The same formula can also be used to calculate amplitudes of rising or setting stars although it should not be used in high latitudes.

LOSS OF THE COMPUTER

If you have been using a computer for working sights you may wonder how anyone ever managed without one, but of course tabular methods have been in use for generations and are the obvious alternative. In addition to the ephemeris, Reeds' and Macmillan's almanacs both include their own sight reduction tables. Each has its own peculiarities and will require practice before it can be considered a viable alternative.

Calculators of the 'scientific' type with sine, cosine and tangent functions are fairly inexpensive and can provide a computer alternative, albeit somewhat slower. With the exception of the moon algorithm, the methods used in this book were chosen because they could be solved in this way but you will need to take methodical care if mistakes are to be avoided. With this in mind it is a good idea to prepare a waterproof sheet with the necessary formulae and proforma to help work the methods with the particular calculator you have aboard. In an emergency you might also consider using trigonometric and log tables, but if they are the usual four-figure tables you will need carefully to consider how rounding off will affect the accuracy of methods normally worked to 8 or more decimal places.

Sight Planning
The star finder (NP 323) described in the last section can be used to predict the positions of all objects and is an especially useful item to have in the emergency navigation package.

High Altitude Sights
With no computer reduction tables or calculator, position lines may still be obtained from sights of objects within a few degrees of the zenith.

The idea of geographical positions was introduced in Chapter 1. Look back at Fig. 1.1 and see how Eratosthenes would have measured the same sun angle were he at any point 575 miles distant from the well, provided that he took the measurement at the time of midsummer noon in Alexandria or Syene when the sun's geographical position was directly on the well. Almanac ephemeris is simply a list of geographical positions of nautical objects for any time throughout the year and these can be plotted directly upon the chart using DEC instead of LAT and converting GHA to LNG as follows:

If GHA < 180 degrees then LNG = DEC and LNG is westerly
If GHA > 180 degrees then LNG = 360 – DEC and LNG is easterly

In practice the method involves using the sextant to take a conventional sight of the object. The altitude is corrected as usual and then subtracted from 90 degrees. This gives the zenith distance of the object and is equivalent to the radius of the position circle. Convert the angle to nautical miles by expressing it in minutes and draw the circle on the chart centred on the geographical position. Whilst the method is certainly quick, inaccuracies arise because it is difficult to determine the azimuth of objects within a few degrees of the zenith.

Meridian Altitudes
Polaris. One of the least complicated methods for finding lati-

The plot was obtained from the following readings:

Time	Sextant Reading
11-42'-30"	53°-40.5'
11-52'-41"	54°-02.2'
11-59'-31"	54°-16.0'
12-04'-03"	54°-24.5'
12-10'-03"	54°-29.5'
12-21'-30"	54°-28.8'
12-35'-49"	54°-10.5'
12-47'-20"	53°-46.5'
12-55'-01"	53°-25.0'

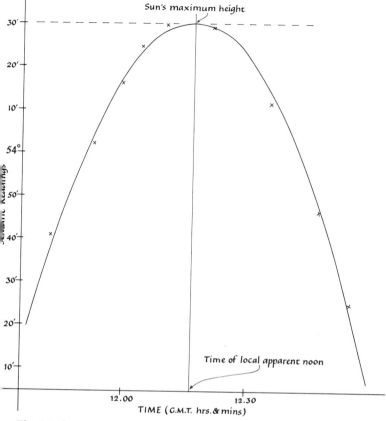

Fig. 8.1 Plotting time against sextant altitude makes it easier to estimate The Sun's maximum height and the time of local noon

tudes in northern hemisphere sights is to use sights of the pole star. Polaris is always on the meridian and its height above the horizon is roughly equal to the observer's longitude. Because the geographical position of Polaris is not located exactly on the north pole such sights can be in error by as much as one degree but the nautical almanac gives sets of corrections for improving the accuracy.

The sun. The procedure for obtaining latitude by measuring the altitude of the sun (or other object) as it passes the meridian has been a cornerstone of astro navigation for generations. No specific meridian altitude methods have been included with the programs given earlier since such sights can simply be run as any other sight. In these cases the computer will print out an azimuth close to zero or to 180 degrees and the latitude is found by using the intercept to correct the dead reckoning latitude.

The reason that meridian sights have been so popular in the past is at least partly because they are easy to work as the only ephemeris needed is DEC. This varies between 23.4 north on 21st June (midsummer) and 23.4 south on 21st December (midwinter). At these times the daily changes are quite small and it would be quite acceptable to use the figure from the day before. Greatest rates of change are about 24 minutes per day and these occur at the times of the equinoxes on 21st March and 23rd September.

Taking a meridian sight of the sun involves measuring its altitude at apparent local noon, the time when it reaches its highest position in the sky and is exactly half way between sunrise and sunset. In practice, it is a good idea to begin taking sights well before local noon and to plot the results on a graph of time against altitude (see Fig. 8.1 for an example). In this way you will at least be able to make a reasonable estimate of the time and maximum altitude of the sun should it cloud over at that particular moment.

Once the altitude has been obtained it is corrected for dip, refraction and semi-diameter in the usual way. Then the zenith distance is found by subtracting the corrected altitude from 90 degrees and this is used to calculate latitude from the following formula:

$$LAT = DEC - \text{zenith distance}$$

In this equation zenith distance is a positive quantity if the sun is observed to the north, or negative if it is to the south. DEC and LAT also follow the usual conventions, being negative if southerly.

With the graphical estimation of the time of local noon, the same sight can be used to give a longitude position. This is because your longitude is equivalent to the difference between the time of your local noon and that of Greenwich. You will need a nautical almanac to find the time of local noon at Greenwich and this can be worked from the times of sunrise and sunset, or from the ephemeris (ie half way between sunrise and sunset, or the time when GHA sun = 0).

In converting the time difference to longitude, multiply hours and fractional hours by 15. This is named westerly if your local noon is later than that at Greenwich or easterly if it is earlier. The main limitation on the accuracy of this procedure is determined by how accurately you were able to find the time of local noon. Although the graphical method is helpful it is not particularly easy, as the sun may appear to hang in the sky at its maximum altitude for some while. A 4-second timing error will lead to a 1-minute error in longitude.

LOSS OF THE SEXTANT

Rough tools for measuring angles are not too difficult to cobble together using scrap or spare parts and a little ingenuity. You may not be able to match the accuracy of a sextant but precision to within one degree is possible. For example, a compass rose cut from an old chart could form the basis of a makeshift quadrant. Also a hacksaw blade can be bent to form an arc where the teeth form regular angular subdivisions, though they will need calibrating in degrees.

Consider also the possibility of using the width of fingers, or a hand held at arm's length. Most people's middle fingers are around 2 degrees in width when held at arm's length, but it is of course best if you calibrate your own limbs before the sextant goes missing. As an example, and in the absence of anything better, the formula for measuring the distance of coastal features of known height given in Fig. 4.1 could be replaced by the following:

Approximate distance off in miles = Height of the object in metres/60 × Number of finger widths

Horizon sights
If you have no sextant then the one time when you will know the height of an object with some certainty is when it touches the horizon and its altitude is zero. The exceptions are the sun and moon, when the altitude will be equal to plus or minus the semi-diameter depending on whether it is the lower or upper limb that touches down.

Record the time of touching the horizon to the nearest second. Although this leads to a negative computed altitude the calculations follow through as normal. You may even be able to carry out a transferred position line fix between sunrise and sunset, and on rare occasions the brighter stars or planets may also be visible at this low angle. The main uncertainty with such sights is due to refraction effects, which will be at a maximum, although the refraction formula in Chapter 4 has produced good results in practice.

LOSS OF THE WATCH
Without a watch you may still be able to use the meridian altitude method to find your latitude, although you will still need to know the date to obtain an accurate DEC from the almanac. However, loss of the watch is serious, for without time you are unable to fix longitude. But is it that the watch is unusable or is it simply that GMT has been lost? Regaining GMT should be possible if you have a radio capable of receiving broadcast transmissions or if you can establish a VHF or visual contact with a passing vessel.

Failing this it may be possible to recover GMT using a lunar distance method or by taking a series of star and moon sights in quick succession. The technique here is to try to get a meridian altitude sight to fix your latitude and then use your best estimated position and time to reduce and plot the results as normal. Because of the GMT inaccuracy it is unlikely that the intersections of all position lines will appear close to each other, so try assuming a watch error and plotting the lines again. This may or may not improve the situation but more trial and error plots should eventually give closer intersections and so provide a fix together with an estimate of the watch error.

LOSS OF EVERYTHING

Emergency navigation involves making the best use of what you have available, and apart from astro methods this could well include taking notice of floating garbage, birds, and passing aircraft as a guide to compass direction. Following any disaster, difficulties are much reduced if you have already been keeping a regularly updated deck log. Not only will it provide the best indication of your present position but it should include navigational trends prior to the event. Tidal streams, differences between your dead reckoning and fix positions, and watch errors are bits of information that assume a much greater importance when you have little else.

In an emergency total loss of everything is unlikely if you have prepared for the possibility. However, following a major disaster, morale is likely to increase if you are able to take an active part in your recovery. You may be able to use some of the techniques given as possibilities following the loss of a single piece of equipment, but your greatest navigational asset will be your ability to maintain a dead reckoning position. To do this you must be able to make reasonable estimates of your speed through the water and your course.

Without instruments, opportunities for obtaining good fixes by astro navigation are rare, but chances for getting accurate course references occur frequently. Making the most of these requires an implicit familiarity with the sky. Getting to know your way among the movements of the sun, moon and planets is not difficult: it is achieved by looking at the sky frequently, and taking an active interest in how the passage of different objects would be seen from other parts of the world. As well as providing an interesting study ashore, this activity has important practical applications at sea and may even, one day, save your life.

APPENDIX 1
Commands used in other forms of BASIC

The following commands have not been used in any of the programs listed in this book, though they may be useful if they are acceptable to your particular computer.

DEG and DMS

DEG is a command used to convert degrees, minutes and seconds into degrees and decimal fractions of a degree. The inverse function is DMS which returns degrees, minutes and seconds.

DEG and RAD

DEG can have a quite different meaning in personal computers which work in radians. Here it is an instruction which returns the value in degrees of an angle expressed in radians. RAD is the inverse of this function, performing the conversion of degrees to radians.

These commands will be needed if you are using a computer that will work in radians but not degrees. All programs listed here are written for degree working, but if you have to work in radians, you will need to change degrees to radians on entry and then back to degrees on exit from the trig routines. The simplest way of doing this is to try making all SIN, COS and TAN statements read as SINRAD, COSRAD and TANRAD. ASN, ACS and ATN would similarly need changing to ASNDEG, ACSDEG and ATNDEG. This approach to the problem could mean that your computer is having to make many degree/radian conversions and accuracy may suffer. In this case a better approach would be to rewrite the listings to work in radians.

DEF and FN

Some versions of basic use these instructions to enable users to DEFine their own FuNctions. A useful example would be to set up functions to carry out radian to degrees and degrees to radian conversions, but you will need to check with the manual for the particular syntax required.

APPENDIX 2
Annual Updates

Before the ephemeris calculations can be carried out variables N, P, Q and R must be loaded with the appropriate annual updates as given in the list below. N is required for all ephemeris, P and Q for the sun and R, the Aries correction, is needed for stars as well as the moon.

N (year)	P	Q	R
1988	−3.8470	77.2633	98.8897
1989	−3.1157	77.2461	99.6382
1990	−3.3702	77.2289	99.4008
1991	−3.6251	77.2118	99.1631
1992	−3.8804	77.1946	98.9250
1993	−3.1507	77.1774	99.6719
1994	−3.4071	77.1603	99.4326
1995	−3.6640	77.1431	99.1929
1996	−3.9212	77.1260	98.9529
1997	−3.1930	77.1087	99.6982
1998	−3.4505	77.0916	99.4579
1999	−3.7078	77.0744	99.2177

N (year)	P	Q	R
2000	−3.9523	77.0605	98.9817
2001	−3.2217	77.0411	99.7300
2002	−3.4828	77.0236	99.4887
2003	−3.7354	77.0092	99.2517
2004	−3.9908	76.9909	99.0146
2005	−3.2630	76.9706	99.7608
2006	−3.5156	76.9577	99.5237
2007	−3.7738	76.9424	99.2783
2008	−4.0293	76.9231	99.0413
2009	−3.3014	76.9041	99.7871
2010	−3.5568	76.8858	99.5504

The updates can be loaded manually. With most computers an example of the usual procedure is to type in, say, N = 1989 then press the ENTER button.

APPENDIX 3
Allocation of Variables

A   A menu variable given the following values:
Sun sight = 1: Star sight = 2: Moon sight = 3: Planet sight = 4. For sight plans A is made negative and this has the effect of omitting the sextant routine and printing out an azimuth and computed altitude.

B   GMT time and date expressed in fractional days plus whole days counted from day 0 at the beginning of the year.

C   Dead reckoning latitude. Later becomes fix latitude.

D   Dead reckoning longitude. Later becomes fix longitude.

E   Declination.

F   Hour angle.

G   Computed azimuth.

H   Sextant altitude. Later becomes corrected altitude.

I   Computed altitude.

J   In computing fixes from paired sights, J holds the azimuth of the previous sight.

K   Holds the intercept of the previous sight (after correction, if necessary, for a change of position).

L   Carries the number of days and fractional days between day 0 of 1980 and the time of the sight. In line 1000 it is changed to the time variable used throughout the moon algorithm.

M   Unused spare.

N   Year.

O   Unused spare.

P   Annual update coefficient for the sun.

Q   Annual update coefficient for the sun.

R   Annual update coefficient for GHA Aries.

S   Semi-diameter.

T   T< >0 inhibits course and distance requests made in fixing positions from paired fixes.

U   A general purpose variable for computation intermediates.

V   A general purpose variable for computation intermediates.

W   A general purpose variable for computation intermediates.

X   A general purpose variable used for character strings.

Y   A general purpose variable used for character strings.

Z   Subroutine exchange and general variable.

Z(1) to Z(5)   *Compact Data* GHA–GMT coefficients A0 to A5.

Z(6) to Z(10)   *Compact Data* DEC coefficients A0 to A5.

APPENDIX 4
Subroutine Mapping

Subroutine number	Description	Start line
(2)	Sextant angle corrections	7850
(3)	Degrees, minutes and seconds to decimal degrees	600
(5)	LAT and LNG loading	500
(6)	GHA and DEC loading	800
(7)	Removing excess 360 degree multiples	7000
(8)	Sight reduction	7800
(9)	Greenwich date and time	700
(10)	Aries GHA	930
(11)	Star DEC and SHA loading	900
(12)	Sun DEC and GHA	850
(13)	Moon DEC and GHA	1000
(14a)	*Compact Data* coefficients, loading routine	9000
(14b)	*Compact Data* method for planet DEC and GHA	2000
(16)	LAT and LNG Fix	5000

APPENDIX 5
Program Testing –
Getting the Bugs out

Many of the mistakes you could make in typing in a new program will be syntax errors. These cause the program to stop and the computer to display an error message giving an indication of the nature and location of the error. These are generally

Appendix 5 Table of examples

	Which sight?	DR Lat Deg.	Min.	N/S	DR Long Deg.	Min.	E/W	Month	Day	Hrs.	Min.	Sec.	SHA/GHA Deg.	Min.	Dec Deg.	Min.	N/S	Sext Deg.	Min.	Index error	N/F	Ht of eye	Semi-dia.	U/L	Any other corr.	Altitude ° ′	Intercept miles	Azimuth °	DEC E	LHA F
1	1	50	10	N	001	20	W	8	23	14	28	20						41	15	2	F	2		L			5 away	229	11.24	35.12
2	1P	50	10	N	001	20	W	8	23	14	28	20														41-34		229	11.24	35.12
3	2	49	32	N	165	12	W	5	2	07	11	21	281	3	45	59	N	36	14	3	N	1					16 away	301	45.98	84.10
4	2P	49	32	N	165	12	W	5	2	07	11	21	281	3	45	59	N									36-23		301	45.98	84.10
5	3	09	45	S	052	20	E						243	40	27	31	S	26	58	1	F	3	15.8	U			1.2 towards	116	−27.52	296
6	3P	09	45	S	052	20	E						243	40	27	31	S									27-28		116	−27.52	296
7	3	28	22	S	062	20	E	12	28	01	0	05						45	43	2	N	2		U			13.8 away	331	11.03	19.68
8	3P	28	22	S	062	20	E	12	28	01	0	05														46-14		331	11.03	19.68
9	4	25	22	S	172	28	W						204	57	04	34	N	46	45	4	N	4			2		6.9 towards	309	4.57	32.48
10	4P	25	22	S	172	28	W						204	57	04	34	N									46-31		309	4.57	32.48
11	1	49	30	N	005	00	W	4	20	13	25	23						49	30	1	F	2		L			3.5 away	206	11.71	16.63
12	1P	49	30	N	005	00	W	4	20	13	25	23														49-47		206	11.71	16.63
13	2	49	30	N	005	00	W	4	20	20	15	09	350	01	56	29	N	21	55	1	F	2					16.4 towards	337	56.48	137.93
14	2P	49	30	N	005	00	W	4	20	20	15	09	350	01	56	29	N									21-34		337	56.48	137.93

NB The year taken for the above sights is 1988. Star used in sight **3** is Capella. Star used in sight **13** is Schedar.

Fixes from Paired Sights.

Position line **11** transferred 117° and **15** miles intersects with position line **13** at 49°41′N 5°24′W.

fairly easy to correct as the computer manual should give precise details of what is required.

More insidious are the kind of mistakes that allow the program to run through as normal and to produce wrong results. Running through the examples in the following table should provide a preliminary check for these.

These examples will pick up most mistakes but are not guaranteed to pick up every possible kind.

With programs that contain ephemeris a reassuring check is to try comparing the computed results with nautical almanac ephemeris. This can be done by inspecting the values of E and F after running through any sight when they will remain loaded with DEC and LHA. Working sights with longitude entered as zero will leave F loaded with GHA which can be checked against the almanac.

Most computers have a facility that allows variables to be checked between each program step. Here you can carry out a methodical and systematic check by comparing data produced at each stage against that worked by an independent method.

Confidence in pocket computer astro navigation comes quickly, especially when you have successfully worked through these examples and tests and used the system at sea for routine navigation. However, do not allow time and familiarity to dull your understanding of the principles on which you are depending to obtain those position lines and fixes. Your best insurance against equipment errors lies in actively thinking about the results you expect to obtain before you get them, then accounting for the differences.

Conclusion

Today's electronic position fixing systems have led many yachtsmen to reject astro navigation. I would imagine also that with the promised accuracy and availability of the coming generation of satellite navigators, many others will be encouraged to use them in coastal navigation, where previously they would have used dead reckoning methods. However, I believe there will always be strong grounds for maintaining independent methods like dead reckoning and astro navigation. In Gibraltar, at least, I have noticed that most people attending my courses are individuals who have already installed satellite navigators. This need for an independent system is also reflected in current RYA courses, where use of electronic position fixing systems is taught alongside traditional methods. A recent development here is that programmable calculators are now permitted for sight reduction during the Yachtmaster Ocean shorebased course examination.

In summary, I feel that learning to program pocket computers will provide a natural progression to the tabular methods taught at present. In an emergency and with the help of a prepared sight form the computer methods can still be worked a step at a time by alternative means. People who routinely use satellite or other forms of electronic navigation will always have a use for astro navigation as a standby system. Those looking for an understandable, reliable and economic system will continue to use it as their main method for fixing mid-ocean positions.

Bibliography

Per Collinder, *The History of Marine Navigation* (Batsford, 1954) (OP)

E. G. R. Taylor, *The Haven-finding Art* (Hollis & Carter, 1956) (OP)

B. Reffin Smith & L. Watts, *Better BASIC, A beginner's guide to writing programs* (Usborne computer books, 1983)

H. Shufeldt & K. Newcomer, *The Calculator Afloat* (Adlard Coles, 1984) (OP)

D. Burch, *Emergency Navigation* (Ashford Press Publishing, UK, 1985)

P. Duffett-Smith, *Practical Astronomy with your Calculator* (Cambridge University Press, 1981)

P. Duffett-Smith, *Astronomy with your Personal Computer* (Cambridge University Press, 1986)

B. D. Yallop & C. Y. Hohenkerk, *Compact Data for Navigation and Astronomy* (London: Her Majesty's Stationery Office prior to 1991) (Cambridge University Press 1991–6)

Reeds Nautical Almanac (Thomas Reed Publications Ltd, annual)

The Macmillan & Silk Cut Nautical Almanac (Macmillan Publishers Ltd, annual)

The Nautical Almanac (Her Majesty's Nautical Almanac Office, annual)

NP 283 Volume 5: Admiralty List of Radio Signals (The Hydrographer of the Navy, annual)

Standard times and radio time signals (The Hydrographer of the Navy, annual)

NP 323: Star Finder and Identifier (British Admiralty publication) (American equivalent 2102–D)